Running After Paradise

Critical Green Engagements

Investigating the Green Economy and Its Alternatives

Jim Igoe, Melissa Checker, Molly Doane, Tracey Heatherington, Jose Martinez-Reyes, and Mary Mostafanezhad
SERIES EDITORS

Running After Paradise

Hope, Survival, and Activism in Brazil's Atlantic Forest

Colleen M. Scanlan Lyons

THE UNIVERSITY OF
ARIZONA PRESS

TUCSON

The University of Arizona Press
www.uapress.arizona.edu

We respectfully acknowledge the University of Arizona is on the land and territories of Indigenous peoples. Today, Arizona is home to twenty-two federally recognized tribes, with Tucson being home to the O'odham and the Yaqui. Committed to diversity and inclusion, the University strives to build sustainable relationships with sovereign Native Nations and Indigenous communities through education offerings, partnerships, and community service.

ISBN-13: 978-0-8165-4013-6 (hardcover)

Cover design by Leigh McDonald
Cover photograph by Rayna Benzeev

Library of Congress Cataloging-in-Publication Data
Names: Scanlan Lyons, Colleen M., 1968– author.
Title: Running after paradise : hope, survival, and activism in Brazil's Atlantic forest / Colleen M. Scanlan Lyons.
Other titles: Critical green engagements.
Description: Tucson, Arizona. : The University of Arizona Press, www.uapress.arizona.edu, 2022. | Series: Critical green engagements: investigating the green economy and its alternatives | Includes bibliographical references and index. | English and some Portuguese.
Identifiers: LCCN 2021053044 | ISBN 9780816540136 (hardcover)
Subjects: LCSH: Environmental management—Brazil—Bahia (State) | Environmental sociology—Brazil—Bahia (State) | Sustainable forestry—Brazil—Bahia (State) | Bahia (Brazil : State)—Environmental conditions. | Bahia (Brazil : State)—Social conditions.
Classification: LCC GE320.B6 S33 2022 | DDC 333.720981/42—dc23/eng/20220104
LC record available at https://lccn.loc.gov/2021053044

Printed in the United States of America
♾ This paper meets the requirements of ANSI/NISO Z39.48-1992 (Permanence of Paper).

This book is written in memory and celebration of the life of Aluisio Medeiros da Rosa Borges (1943–2019)—professor, development specialist, deep thinker, and dear friend who set me on my path in life nearly three decades ago by posing a question:

"Why don't you go to Brazil?"

There are people in life who watch things happen.
There are people in life who make things happen.
And there are people in life who ask "what happened?"

This book is dedicated to those in the middle, to those who "make things happen" and, in doing so, who sit at the intersection of ideas, places, cultures, purposes. There are many people in my life who helped to make this book happen and I dedicate this to them.

To my mother, who makes things happen for countless others in countless ways every day.

To my "Little Family" of Jeff, Max, Ella, and Maya, who make things happen in Boulder, in Bahia, and in the places in between as we explore and love our life together.

To the people featured prominently in the pages to follow: Maria Guiomar Carvalho Silva, Rui Barbosa da Rocha, Maisa Fontana Amaral Silva, Maria do Carmo Tourinho Nunes, Alba Valeria Nunes de Melo, Jitilene Silva dos Santos, Otília Nogueira, Erasmo Carlos dos Santos Cruz, Miguel Carlos Costa ("Catu"), Arionilton Soares de Sa ("Comprido"), Eliel Roberto dos Santos ("Irado"), and Maria do Socorro Ferreira de Mendonça. Each of them has generously opened their lives, their stories, their efforts, and their dreams to me as they run after, as they make things happen, in Southern Bahia.

Contents

Illustrations

Figures

Maps

Acknowledgments

While people often characterize writing as a solitary endeavor, I believe it is done in community. Though an author is ultimately the one to put words upon paper, this takes place in the company of those who serve as inspiration and impetus, as subject and object of the stories we tell and the deeper messages we seek to convey. In writing, and in life, I adhere to the advice of Catu, one of the people featured in this book, who once remarked, "I try to surround myself with good people." There are countless good people who have encircled me and helped me create this book in one way or another.

The ideas, themes, and stories stem from my dissertation in cultural anthropology, which was guided by some of the best academic mentors a person could ask for: Emily Yeh, Kaifa Roland, John Collins, Carole McGranahan, and my primary advisor, the inimitable Donna Goldstein. I am grateful to each for contributing to my work while inspiring me with theirs. In particular, I am continually heartened by John's humor coupled with a deep understanding of the complexities of Bahia. I am inspired by Carole, my writing and thinking muse, who has steadfastly encouraged me to write what I want, *as* I want, while modeling what this practice looks like herself. From the moment Donna and I met, "*agente bateu forte,*" we clicked. I am ever grateful for her wisdom, generosity, and deft insight into my work for nearly two decades and honored to be in the company of a master who taught me the importance of engaged, theoretical storytelling about the complex and fascinating place that is Brazil.

My Colorado-based community of writers, thinkers, and friends has buoyed me throughout this endeavor in many ways, often led by a band of anthropologists. Carol Conzelman and Laura Deluca provided important insights on this manuscript and fervently cheered me on as I worked to get these ideas out into the world. Terry and Judith McCabe provided ongoing support for this project by serving as a solid sounding board and encouraging me professionally, which, in turn, gave me the conviction to write. Without the dynamic duo of Nicole Smith and Alicia Davis, my dissertation never would have been completed, and, in turn, the seeds of this book would not have been sown as we contemplated, critiqued, wrote, edited, cried, rewrote, drank tea, laughed, and advanced our work together over countless kitchen table gatherings. I also appreciate the insights and encouragement of my valued colleagues in the Environmental Studies Department at the University of Colorado, namely Joel Hartter, Peter Newton, and Rayna Benzeev, who give me continual reminders that collaborative fieldwork, research, and teaching make life not only more productive but also more fun. I particularly thank Pete for his keen sense of maps and Rayna for producing a map that tells a particular story and for providing a brilliant cover photo. I am ever grateful to my broad circle of friends in Boulder and beyond who supported me by asking (for years), "how's the book coming?" This simple question revealed to me how much they cared and gave me the push I often needed to continue along my writing pathway.

I am thankful for and indebted to my "creative comrades in writing" in Boulder, Matt Moseley and Brook Eddy, both of whom read this work cover to cover and offered astute insights on tone, voice, story, flow, and, perhaps most of all, provided brutal honesty on "what to cut." Matt and Brook's personal creative work, united with a sense of adventure and keen reading of the world around us, inspires me to be better in my own commitment to the craft of writing. Rosalind Wiseman, one of the most experienced "good writers" I know, challenged me to think about agency and identity just as this project neared completion; thanks to Ros, the people in this book have their real names and, in turn, their voices are louder. Linda Tate served as an editor extraordinaire by understanding and honoring my efforts to walk the fine line of theoretical storytelling; she steered me back on course when I strayed a bit too far and continually believed in this project, even when my own faith in it wavered.

I am fortunate to have *really* good family in my community of good people. I am grateful to each of my siblings, Martin, Daniel, Molly, Sam, Eddie, and Kay, and my parents, Edward and Sheila, who supported me in spirit throughout this process as well as to my "family members who write" who went beyond. Martin provided regular counsel on the writing process, the importance of deadlines and accountability, and the need to "celebrate the small milestones" along the winding pathway toward a book. My sister-in-law, Melissa Kwaterski Scanlan, inspired me by her own work on cooperatives and collaborative engagement, themes that align with this book, while Molly Cummings, my "near sister" and yet another tenacious scholar in her own right, cheered my writing process with the boundless enthusiasm that only she possesses.

I offer heartfelt gratitude to the editorial community at the University of Arizona Press, in particular the editors of the Critical Green Engagements Series. Jim Igoe, Jose Martinez-Reyes, and Allyson Carter believed in and supported this project from the moment I declared "I want to write a different sort of book" to the time of its completion. Two anonymous reviewers made this manuscript infinitely better with thoughtful reading and pointed comments. I also thank all members of the team at the University of Arizona Press, Amanda Krause, Leigh McDonald, Abby Mogollon, and the others who deftly guided me on the long and winding road of initial concept to final production.

Colleagues at the State University of Santa Cruz (UESC), namely Salvador dal Pozzo Trevizan, Romari Martinez, and Deborah Faria have been, and still are, invaluable collaborators in my research, writing, and teaching in Southern Bahia. Doctoral student Carolina Menna generously helped with the translation of a draft of this manuscript, which allowed for those featured in it to review and speak back to my interpretations. Photographer Jose Nazal offered valuable insights on the regional maps and photos.

Finally, the good people in the Southern Bahia region are those who truly made this book happen, from the initial ideas for its inception to the ways in which it will go forth in the future. Some are unnamed and entered my life through fleeting, yet significant, acts of kindness and generosity that made my work not only possible but joyful. Others, named explicitly in the book's dedication, were—and still are—my teachers. I am grateful to each of them in different, yet interconnected, ways. I thank Maria Guiomar Carvalho Silva for embracing me as *família* and serving as a co-mother to my children and a sister to me. I am grateful for the intention and thoughtful curiosity that Rui Barbosa da Rocha puts forth in the world in virtually everything he does, and for having the generosity and confidence in me to share so much wisdom, from the philosophical to the practical. I am continually inspired by the tenacity, commitment, and "can do" spirit of Maisa Fontana Amaral Silva, who has never stopped fighting for her community, as well as by the thoughtful insights, fearless determination, and unstoppable attitude that fuels the daily work of Maria do Carmo Tourinho Nunes and Alba Valeria Nunes de Melo. Jitilene Silva dos Santos and Otília Nogueira have always welcomed me into their homes with open arms and, through this small but profound act, entrusted me with their stories, and their lives, in profound ways. Erasmo Carlos dos Santos Cruz inspires me with his commitment not only to the wonders of Southern Bahia but the beauty of sharing this with others. Miguel Carlos Costa, "Catu," has continually kept me positive by his upbeat spirit and unwavering belief in the power of good people working for a better world. Arionilton Soares de Sa, "Comprido," has shown me that sharing movement, story, music, and art with youth is one of the most powerful ways of keeping culture alive. Eliel Roberto dos Santos, "Irado," has provided continual comic relief through the trials of life, which were many, in Southern Bahia. And Maria do Socorro Ferreira de Mendonça has shown me the power of being rooted in place and how one person can, in turn, inspire others to tap into and mobilize their convictions to fuel activism throughout an entire region.

My community of Brazilian friends and family that spans well beyond Bahia has supported me in my many, many journeys into Brazil both literal and figurative. Mario Mantovani and Paula Arantes inspire me not only with their commitment to social-environmental activism but also with their boundless enthusiasm for this important work. And, for nearly three decades now, my first Brazilian family, the powerful (and predominantly female!) Guerras, in particular my Brazilian "sister" and "son," Cecilia and Pedro, have continually welcomed me into the open, intellectually engaged, and sometimes raucous fold that is distinctly theirs. Without their love and honest tutelage, I never would have come to know Brazil as I do.

A community of good people is made real through particular places and practices, and I am grateful to the environments that existed, and that I could create, to find the presence and inspiration I needed to write. I appreciate the St. John's Abbey Guesthouse in Collegeville, Minnesota, which served as an ideal writing retreat when I needed it most, and the music of Caetano Veloso, Gilberto Gil, Tom Jobim, Nati-Roots, and Falamansa, which fueled my soul through many an hour of thinking and writing and rewriting.

Finally, I am truly blessed that some of the very best good people in my life happen to be my immediate family. I have a deep gratitude to and appreciation of my three children—Max, Ella, and Maya—for embracing (and sometimes enduring) my fascination with all things Brazilian. Being raised by a mother who does fieldwork far from home, spends early morning hours and weekends writing, and constantly challenges the family to bridge the divide between Brazil and Boulder by immersing them in a steady stream of Brazilian music and friends from afar in our home could not have always been easy. They have not only tolerated this lifestyle but usually embraced it, even when they were forced out of their comfort zone in all sorts of ways.

And the foundational core of my circle of good people is my partner in life, Jeff (aka Jefe). It is both inspirational and daunting to be married to the best writer one knows. Jefe read every sentence in this book (some way too many times) and spent countless hours reflecting and offering honest and constructive insights in ways that only someone who has lived these experiences with me can. I appreciate his constant mantra: "include more stories of people!" The interesting parts of this book I owe to him; the less interesting ones I must take full responsibility for. From the day long ago in 2003 when he stood on a small beach in Southern Bahia, coconut in one hand and beer in another declaring "this is a pretty good place, you might want to consider working here," Jefe has endured the lowest of lows and celebrated the highest of highs in my life. I thank him for loving me through it all. He, and the rest of our "Little Family," are the best traveling companions—both literally and figuratively—that I could ask for in this life.

Running After Paradise

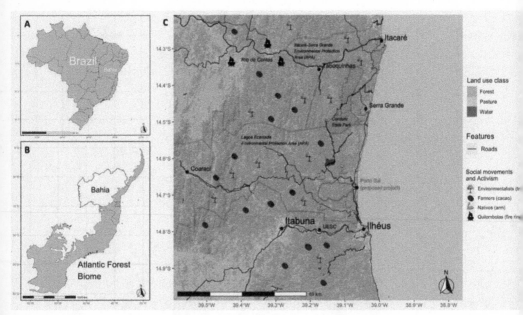

Map 1 Southern Bahia region within Brazil, within Bahia, and the diverse social movement actors (marked by symbols and locations) discussed throughout this book. Map by Rayna Benzeev.

Introduction

Crafting the Stories, and the Story, of Here

"Each little region along this coast has its story."

—Rui, Ilhéus, Bahia

A dozen hands reach up, while one reaches down. Shrouded in a hoodie and *capoeira* pants, carrying a single duffle and a *berimbau*, Neto has just hugged his mom goodbye, choked back tears, and slowly walked up the stairs of the bus to his window seat.[1] He squeezes his brawny arm through the grimy, narrow bus window for one last touch of the outstretched fingers trying to grasp his hand. Perhaps these points of contact remind him of where he was coming from, for he knows, as do I, he has no idea where he is going. I press a 50 reais bill, about US$25, into his outstretched palm. I know it is a minuscule contribution for all that lies ahead. I also know this is a lot of money for

---

1. If you have dropped down here to the footnotes, Dear Reader, you will find that herein lie other stories that further enhance the stories above. Sometimes these "stories below the stories" introduce historical facts or complementary information. Other times, they draw extensively on others who have thought these things through in similar, or different, ways. In similar, or different, places. Enter deeply, superficially, or not at all to this "guide below the guide" to life in Southern Bahia, Brazil.

Capoeira is a martial art that arose in resistance to slavery in Brazil and is now practiced throughout Brazil and the world. Thompson (2000, 103) explains, "The actual word capoeira means 'brushwood' or the long waist-high grasses enslaved people hid in when they had escaped. The swaying, swinging, *ginga* movement positions this activity as both dance and struggle, political and cultural, real and rehearsal." The *berimbau*, made with a long bowed stick, wire, and gourd is a key instrument, a prized possession of capoeiristas. See Downey (2005) for a rich anthropological analysis of capoeira.

him. The bus starts its engines loudly. This is the last chance to turn back and jump off, but Neto is already mentally gone. He is ready to head *la fora*, or "out there," as the surfers from his town call the waters beyond the wave break. Getting out there requires tenacity or *força*, force, as they say in Portuguese. It also requires faith that life will be good once you get where you're going. Neto has these things. His mom wipes her tears. She knows he is ready to go and his options are limited here.

Music streams from a nearby bar as men drink from a communal bottle in a beat-up coozie. They pour small glasses for each of them, then drink and slap dominoes on a plastic table perched precariously on the cobblestone street. The bus backs out of the station and slowly lumbers up the big hill leading out of the small surf town of Itacaré, where Neto was raised. The Atlantic Forest engulfs the road as the ocean far below transforms into an increasingly distant mirage. Neto's friends and family slowly turn away from the bus station and head back toward their lives. While he is leaving to make a life out there, they are staying to create theirs here.

"Here" is the Atlantic Forest of Southern Bahia, Brazil—an important, and contested, place. The "story of here" is what I came to pursue. Here, people with deeply entrenched differences of history, class, race, and identity must increasingly interact with each other not only because they live in one of the most environmentally threatened areas on the planet but also because they face the collective need for, as well as threat of, development and change. Here, individual and communal dreams, ways of living, and ways of making a living conflict, then find ways of coming together, then conflict again. This rhythm is mirrored in the region's natural environment. Ocean waves roll up to the coastline, crashing audibly and violently, and then travel out again, giving way to a pause of peace, respite, calm, and possibility before they come crashing into shore again. Tropical rains fall, starting almost imperceptibly with a smattering of tiny droplets that ping lightly, gradually and steadily building to a deluge that torrents relentlessly upon the forests, farms, homes, and streets. Until suddenly they stop. Clouds cede to bright sunlight, again, and bathe the region with a new moment of opportunity before the rains return to momentarily halt and then fuel life again here.

This book is about tales of conflict and convergence. It is about a complex and heterogeneous people and landscape that are shaped by realities of the past and dreams of the future. It is about how individual stories comprise a collective story of here, a story that is revealed through cultural politics of social movements that shape this region both on their own and in relationship to each other. It is about the blatantly obvious as well as the delicately nuanced ways in which people are striving, negotiating, crafting, and "running after" life in Brazil's Atlantic Forest.

Arriving

As I fly into Ilhéus, the main point of entry for Southern Bahia, my eyes take in the rolling hills and cleared pasturelands truncated by swaths of dense Atlantic Forest. If I sit on the right side of the plane, I can see the small Cristo, a replica of Rio's iconic Christ statue, in the city's harbor. My heart always skips a beat. The kids often grab my hand and glance knowingly at me. We're back.

Getting off the plane, we descend the stairs. Looking up, I see people on the second floor of the airport welcoming the plane. They press their noses against the glass, standing two deep, looking for a view of the action. The humidity hits us like a wet blanket; the smell of the tropics hangs in the air. There are no airplane walkways here to control the transition from where we were to where we are now. From the moment we land, we must enter fully.

We have ample stories of departure and arrival, of near-missed flights and of children wailing, vomiting, melting down along the way. Sometimes the kids cry for hours. In frustration as well as solidarity, I, too, sometimes cry with them as our small family unit is engulfed in the timeless fog that multiple interminable international flights produce. Stories come and go, surge sporadically in my memory, and then recede for years. There was the time a kind woman aided me when my daughter Maya was an infant, during a particularly difficult leg of the long trip. She had dropped calming herbs, aptly called Rescue Remedy, into Maya's mouth each time she opened it to scream and, in looking at the tears streaming down my face in the utter helplessness of the situation, had kindly observed, "Here, you could use some too," as she deposited them in my mouth and caused me to cry even harder with her kindness. There was the time Max, at the mere age of six, had bravely offered to fly home alone, via Canada, during one of our regular flight mix-ups. There was the time all three children vomited and had diarrhea—at the same time—on a small plane. There was the time we lost our bags and stayed in Rio for three days wearing little more than beach *cangas*, towels, to tour the town. There were bouts after bouts of lice. The list goes on. The kids are older now. They no longer scream but now stare into screens or books they can read on their own. Though I miss the chaos of nearly two decades ago, none of us has waned in our desire to return. We frequently characterize this place as "our other home."

Each time we land, the story of arrival repeats—some version of the same narrative and routine, like the familiar ritual of a Catholic Mass broken only by variation in homily and song. Stumbling into the small airport in Ilhéus, we wait for our bags on the antiquated conveyor belt while I instinctively search the crowd. Standing with signs ready to pick up tourists and bring them to Bahia's most chic destinations, the locals, called *nativos*, who work in the region's burgeoning tourism industry, are easy to spot. They bear the name of the small hotel, or *pousada*, on their shirts.[2] Some look alert and eager to be participants in the tourist economy as they don expressions of seriousness and professionalism despite wearing surf shorts and flip-flops. Others appear bored with the routine of yet another airport run on yet another day with yet another group of tourists.

My person, named Dona Guiomar, is the one I look for upon every arrival. She is often a little bit late. She does her hair for the occasion, donning a tight nylon cap, called a *touca*, the night before our arrival to ensure that she awakens with a smooth mane. She will wear her best city clothes and high heels, rare and impractical in the cobblestoned beach town where she lives

---

2. Because of the importance of nativo identity and "being nativo," I retain the term in Portuguese throughout this book to honor its significance. I am using *nativo*, or native, to describe an entire group of people who proudly call themselves this—not dissimilar to bumper stickers one sees in my home state of Colorado that have the mountains of our state and the word "native," or "from here."

At the same time, and as will be revealed throughout this book, there is another meaning of native within anthropology. Viveiros de Castro (2002) writes on power and the relationship between anthropologists, the anthropological gaze, and "natives," noting: "The 'anthropologist' is someone who talks about the speech of a 'native.' The native does not need to be especially wild, or traditional, nor natural where the anthropologist finds him; the anthropologist does not need to be overly civilized, or modernist, not even foreign to the people about whom he speaks. The speeches, that of the anthropologist and above all that of the native, are not necessarily texts: they are any practices of meaning. The essential thing is that the anthropologist's discourse (the 'observer') establishes a certain relationship with the native's discourse (the 'observed'). This relation is a relation of meaning, or, as it is said when the first discourse intends to Science, a relation of knowledge" (113). This tension, between studying Southern Bahia as an observer or outsider and living through the process as a participant, is a tension all anthropologists face, one that will be made real in the pages to follow.

an hour away but more appropriate for an airport run. At some moment in the chaos of our arrival, our eyes will lock upon the discovery of each other, and she—and we—will beam. The initial greeting has become routine as I push the kids toward her and exclaim, "There she is. See her? Go say hi!" For a few weeks we will be together. While our time apart is bridged by WhatsApp texts, calls on holidays, or transactions when she needs "a little help" for a doctor's exam, an educational expense for her grandson, or a trip to the real hair salon, this fleeting time together in person moors us during the spaces in between.

We will cram children and luggage into our dilapidated white VW bug, or *fusca*, that Guiomar will have paid someone to drive to meet us at the airport. The driver will need to travel home via bus, for we are too many to fit on the return. The fusca, which we fondly call the *fuscinha*, or little fusca, is our literal and figurative form of transportation. It allows us to travel the region creating our stories. Guiomar stores the fuscinha when we are gone. One year it even lived in the main room of her restaurant. She would carefully push the tables back and roll the car inside as night fell, locking the grated door behind. "I need to take good care of this," she rationalized, "and you never know what will happen if it is left on the street."

As we head north along the coast, I will ask about Guiomar's life, her family, the latest gossip, and all that has happened since I last left. Almost every time, she will begin with a simple statement: "*Tô correndo atrás*," meaning "I'm running after it." Running after describes the constant and ceaseless effort that she, and countless others here, engage in every day. "It" doesn't need to be specified really, nor can it be. "It" is both specific and general, concrete and ephemeral. "It" could be financial security, more opportunities for her kids and grandkids, or even a better community with trustworthy politicians, less trash on the street, and a good healthcare and educational system. "It" could be a job for the week, a bag of cement for her house, a new tooth. This running after is a part of the rhythm and routine, the past and potential of living and being alive in Southern Bahia, Brazil.

Guiomar will feed us a steady report of local gossip peppered with facts. She will ask to slow the car at places along the way that mean something to her. That's "my" tree, she will say when she sees a pink flowered ipê tree standing out against the sea of green in the tropical forest in the distance. She will point out roadside stands laden with fruit and praise the riches of the land. She will roll down her window and catcall passersby: "Hey big belly!"

Figure 1 Guiomar proudly shows off the region's famous cacao.

to an ample-sized man teetering on a small bike or "Hello kids!" to thin, barefoot children on their way home from the beach carrying homemade fishing poles and beaten-up boogie boards. I count on her to deliver a stream of chatter and gossip frequently truncated by a loud, cackling laugh. I count on her to dissipate my exhaustion and welcome me back. I count on her, this small fireball and the mother of now long-departed Neto, for everything.

Starting with a Question

I didn't arrive in Southern Bahia by coincidence. This quest, this journey, began decades ago. I remember well trudging through the sloppy streets of Washington, D.C., in a cold, mid-February drizzle. I had traveled here to meet with people working in conservation organizations—the big ones that most people have heard of, Conservation International, the Nature Conservancy, the World Wildlife Fund (WWF), and the International Union for the Conservation of Nature. I even had meetings lined up with scientists from the National Zoo who were running a project in Brazil. I was wide-eyed and well intentioned, determined that these conversations would help me to make my research in Brazil meaningful rather than waste my time on an esoteric project that no one would care about in the end. In the early 2000s, Brazil's domestic environmental movement had been growing steadily stronger with new, locally led nongovernmental organizations (NGOs) cropping up by the day. I knew that this in and of itself was interesting and significant, but I wanted to examine environmentalism from a social lens in some way, and at this point I was trying to discern what threads to follow, what to tease out, what would truly matter.[3] After three long days of informational interviews with generous willing staff from the BINGOs, the big international nongovernmental organizations as they are sometimes called, I was exhausted and, I feared, not much further along in my quest. As I wrapped up an interview with a generous, knowledgeable senior staff person at WWF, she paused to reflect on my parting question to her: "What do *you* need? What can help you in your conservation work in the Atlantic Forest?" She glanced outside, then turned back to look at me squarely. "You know," she said, "we're a conservation organization. But increasingly we need to deal with farmers. They often have land that goes right up to the edges of protected conservation areas." She paused again, as if deciding how much to reveal, then continued, "But our organization wasn't exactly founded to do this. Anything you can tell us about farmers is helpful. We need to understand *people* better." This frank, and seemingly simple observation, set my research for this book in motion.[4]

The offices of WWF weren't the only place that I perceived an underlying tension between the natural and the social, which seemed present in virtually all of the important biomes of Brazil, from the miners in the Amazon region to the agribusiness industry in the Cerrado to the small-scale farmers who lived alongside important protected areas of Brazil's rapidly dwindling Atlantic Forest. I didn't plan for the diversity of cultural politics that would come into play to shape these sometimes nuanced, sometimes overt conflicts

---

3. I have been inspired by Peterson et al. (2010), West (2006), and Brosius, Tsing, and Zerner (2005) in thinking about the complex dynamics and power differentials when looking at environmental conservation through a cultural lens.

4. Doane (2012) begins her book with a tale of WWF as she explores the uneasy alliances between transnational environmentalists and rural producers, including the intersections between agrarian conflict, conservation, and Indigenous social movement leaders.

between conservation and livelihood development, between historical practices and future dreams, between how some people view this region and how
others strive to carve out their particular place in it.[5] I didn't plan for the
complexities and dynamics, the ways of seeing and being that converge, then
diverge, then come together in new ways, flowing in, then out, then back in
with new configurations that are shaped by, but different from, the patterns
of the old. I didn't plan for my own personal involvement as I, too, entered
this rhythm of ebb and flow, stasis and change. But once you meet people
and enter into their lives, it is hard to remain unchanged.

Meeting Guiomar

Dona Guiomar came into my life like a sign from God that I could actually
do this. I remember our first meeting like it was yesterday.

One week into my time in Southern Bahia, all systems have broken down
badly. The house is a disaster. With stagnant pools of water surrounding it
and a steep, open staircase with no railing and a concrete floor below, it is
a breeding ground for malaria and the ideal venue for one or more of my
children to fall to their death. The nanny isn't working out either. While she
was an ideal sitter for us at our home in Boulder, Colorado, she has since
returned to Brazil, given birth to a child who is now nine months old, and
both she and I naïvely think that she can manage to raise her new daughter
as well as my brood all together in precarious conditions. It takes us about
two hours to realize this is a ridiculous plan. Even the school situation isn't
as planned. It is too far for us to walk, we have no car (as this is pre-fuscinha
purchase), and apparently the children aren't even enrolled despite months
of paperwork and diligent faxing in an effort to ensure that exactly what
is happening to us now doesn't happen. Jeff will soon be returning to his
job after five weeks of helping to get our family set up in Brazil, which will

---

5. As an anthropologist I am naturally interested in the concept of culture, "a flexible and
powerful concept that can be used in many different ways including, most importantly, as part
of a political critique" (Ortner 2005, 36). Alvarez, Dagnino, and Escobar assert "all social movements enact a cultural politics," sites of struggle against domination and hegemony and ways in
which subaltern actors can join together and mobilize around different conceptions of meaning
and place ([1998] 2018, 6). Cultural politics are instrumental for understanding themes throughout this book, including power, resistance, and new forms of interconnection across difference.

leave me alone to resolve this debacle on multiple fronts while starting my research. The stress is palpable, and I am panicked.

On a hot Sunday afternoon, I crumple into the Indian pillows that adorn the floor of the only other American, a woman named Serena, in the town we have chosen to live. With her I can speak English. I can complain. Most of all, I can cry. Amid my anguish, Serena tries to counsel me on life here. Jeff looks sympathetic while drinking a beer in the corner, knowing there is not much he can do. "It gets better here," she reassures me. I cry harder, stressed by the research I have to complete. Serena has adeptly figured out how to live life quite well in Bahia. She works four hours at night selling beautiful Indian garments to tourists and hangs out on the beach during the day smoking joints with her Rastafarian boyfriend, Marcos. I am more than a little envious of this lifestyle. Sensing my desperation, Serena offers, "I may know of someone to help you. She's cool. But," she hedges, "I'm not sure if she will work for you or not," giving me a sidelong glance to ascertain my interest and perhaps also assessing if I am a safe bet for sharing this contact of hers. I nod, trying to pull myself together after my breakdown.

The next morning there is a knock on the door of the new house we have miraculously managed to procure and rent. I peer through the glass to find a small woman of forty or so looking in at me expectantly. She has long, wavy, hennaed hair, a wide smile, and a short jean skirt. She enters cautiously, peering about and assessing the situation. "My name is Maria Guiomar, but you can call me Guiomar. And I don't just work for anyone," she states at the outset, clearly sizing me up and establishing this is not to be a patron-client relationship. She continues, "I can't watch three kids and cook and clean all together." I nod and offer, "I'll get another sitter to help with the two older kids." She continues to look at me, gauging what else I will give in to. "I'll get a cleaner," I offer, then speak my piece. "But I do need you to watch the little one, Maya, while you cook," knowing that making a meal can take the better part of the day in Bahia. "She won't be in school, she doesn't talk, she's still nursing. And though she is generally sweet, she can cry like a demon sometimes." I pause, waiting for her reply. It doesn't take long. "I'll do it," she says, looking into my eyes. With this, our arrangement, our unwavering connection, begins.

Guiomar came to Southern Bahia more than two decades ago. "I wanted to leave my family behind," she explains. "They are more conservative than I, you know. They are all *formado*," she often says, meaning they have all gone to college to "become formed," which is a strange way of noting in Portuguese that people have a college education. "I was the wild one who chose to marry a capoeira master, Edmundo, and leave for Bahia." She sighs. "But I have no regrets about coming here."

Her life hasn't been easy, but even so Guiomar has more than most. She has property in one of the region's main towns, Itacaré, as well as a couple of small plots of land outside town. "I need to rejuvenate myself with fresh streams," she says. I find this slightly ironic, this notion of Guiomar as stressed, as the town where she has chosen to make her life is known by Brazilians and foreigners alike as a mellow, restorative place with ample beaches, verdant forest, and an overall relaxed vibe that is truncated only by its vibrant nightlife. Guiomar also has countless friends from her decades here raising her children, creating a life. Some friends are wisely chosen—they give her protection if and when she needs it, and sometimes she does. Most of all, she has grit. "I've survived a lot," she often says. In Portuguese, they call these types of women *guerreiras*, warrior women.

Journeying In

As Guiomar, the kids, and I travel from the airport, the colonial city of Ilhéus, at first glance, has an unassuming, understated tropical beauty. Those passing through can concentrate on the palm trees, the picturesque harbor with people stand-up paddleboarding and drinking beer by the beach, and the subtle chaotic buzz of the city's urban traffic and shopping district. Or they can look beyond this veneer. As well-heeled shoppers race along the downtown area to their jobs in banks or offices, a one-legged "car watcher" tries to sell a freshly caught fish of questionable origin to cars that pass him on the narrow street. Street vendors peddle tropical fruits in wheelbarrows, pausing to offer samples if passersby can look past the dirty cart to the ripe wares atop. Electronics stores advertise refrigerators, washing machines, and stereos that can be bought on monthly payment plans. Music and noise emanate everywhere—from shops, car trunks, and people pedaling bikes with portable speakers perched on their handlebars. Everyone seems to be rushing, calling, selling, cajoling, hustling, and competing for the attention of those walking by. They are all "running after" in their own way while the verdant vegetation of Brazil's Atlantic Forest silently, viscerally looms in the hills beyond the city. Herein lies a land of opportunity, a place of possibility.

We speed along the coastline, past the magnificent Roman Catholic cathedral, past the Vesúvio restaurant with one of Brazil's most famous authors, Jorge Amado, memorialized by a bronze bust in front. We drive past the Bataclan, a famous brothel-turned-chic restaurant. Inside, black-and-white photos of the legendary prostitutes from the turn of the century hang on tall stone walls while hidden tunnels in the rocks behind the restaurant lead to the cathedral nearby. Upstairs lies a museum of this era, when clients could run between church and brothel, hidden from the gaze of the masses, their wives, and anyone else who wanted to expose them.

Farther down the street are stately but now rundown mansions with faded pink and orange façades. Their views over the ocean must be stunning, but most of the windows are shuttered.[6] "Doesn't your friend still live in one of these grand houses?" I ask Guiomar, nodding to the homes. "Yeah, but she is always traveling. She has a good life. She has a lot of money," she states matter-of-factly, and looks out at the ocean as if she, too, may be imagining leaving. Perhaps she is thinking about her son Neto, who got on a bus a few years ago to go la fora, out there, and has yet to return.

On the drive out of town, we stop at the gas station. Plastic tables are set up in front of the small convenience store. Every night people gather here for a beer and a snack. On weekends there may even be a band. I buy a road beer for each of us and cookies and chips for the kids. This is our tradition. Rules here are lax, for all of us.

As we depart the city limits of Ilhéus, the faint aroma of chocolate permeates the air. A quaint Swiss-looking structure called Chocolate Caseira, meaning "homemade chocolate," lies at the edge of town to remind folks they have one last opportunity to purchase the region's riches. But for those in the know, the product is less than high quality and certainly not homemade. Reputedly, the finished chocolate product is actually imported from São Paulo and packaged with the brand of Bahia for the unsuspecting tourists,

---

6. Ilhéus is a colonial city with a rich past and a present-day aura that hints at its bygone days. In 1500, Pedro Álvares Cabral landed 300 kilometers south of the city along what is called Brazil's Costa do Descobrimento, or Discovery Coast, claiming Brazil as a Portuguese colony. The colonial process set the stage for subsequent land appropriation and distribution throughout the region. Grand mansions were built in the high point of the cacao era, which began at the end of the eighteenth century.

or my kids, who clamor for the cheap chocolate every time they see the Chocolate Caseira stores that dot the city. "I want a chocolate turtle," Maya always says when we land at the airport and see the familiar stores. A sugar-infused milk-chocolate turtle somehow always marks this as home, despite the fact that the best cacao beans are too often exported from the region for processing elsewhere.

After we cross a narrow bridge over the Rio Almada, shared by speeding cars as well as bike riders and pedestrians coming from the beach or en route to a fishing spot along the riverbanks, a roundabout marks the journey north. The ocean lies to the east, and the rest of Bahia—and Brazil for that matter—is west. The highway, BA-001, is a two-lane, sixty-kilometer paved road constructed in 1997 during the age of ACM, Antônio Carlos Magalhães, Bahia's longest-serving ruler whose legacy permeates to this day.[7]

Picking up speed along BA-001 allows for smooth travel, save for a few potholes and stray dogs. Though the two-lane road lacks a shoulder on either side, this doesn't deter people from biking, or sitting, along the region's main thoroughfare. A mangrove area along the Rio Almada appears to the west where people fish, collect crabs, have picnics, swim. The official story is that BA-001 was built to attract tourism to the region; today, it leads north along the coast from Ilhéus to the tourist town of Itacaré, the Maraú Peninsula beyond, then the Bahia de Camamu, and, to the north, Bahia's capital city of Salvador. The coastal landscape is ideally suited to both domestic and international tourism, with miles of palm trees lining sandy beaches and smatterings of primary and secondary tropical forest abutting the road as a backdrop for the beach. In fact, the highway from Ilhéus northward has officially been designated a special tourism corridor. Attractive signs hang high near the tree line, publicizing its unique characteristics for drivers and passengers who can lure their eyes away from the pristine coastline.

We pass a sign for a condominium called Light and Ocean. "Flaviana used to have a *casa de praia*, a beach house here," Guiomar notes, pointing in the direction of a simple dirt road that leads east to the coast. These secluded

---

7. Bahia retains aspects of its colonial legacy, as seen in the concentration of power of the ruling elite. ACM was the governor of Bahia 1970–75, 1978–83, and 1990–94 as well as the Minister of Communications 1985–90 and a Brazilian Senator 1994–2007. He is known for *carlismo* (named after him), a "way of doing politics" that is a regional version of a combination of economic modernization and political conservatism. ACM's grandson, ACM Neto, became the mayor of Bahia's capital, Salvador, in 2013 and again in 2017. See Herrmann (2017).

properties, hidden from the road, are only visible when one walks along the endless white sand beach they occupy. They stand silent until they are brought to life by summer holidays, Carnaval, or other vacation times, which, in Bahia, are ample. I nod, thinking of her daughter Flaviana, a beautiful and smart young woman whose ambitions for surfing and for school were thwarted by a young pregnancy and a marriage to a not-so-good guy. I also think about how here, in front of this very bucolic-looking condominium complex, two women were carjacked and murdered a few years ago. Tragic, disconcerting events like this are often hushed, or completely hidden, leaving those in the know with the task of deciding what to pass along and to whom.

Continuing along the highway for a good twenty-five kilometers, we pass small groups of houses, roadside bars, and run-down roadside condominiums. Tattered nets hang above the highway to facilitate animal crossings, biologists' attempts to mitigate the inevitable effects of regional development on the local fauna. We race past an empanada restaurant, a site that marks the thin lines between my life and my work here. Sometimes I am with the children, and the empanada stand is a place to feed them as we break up a long road trip. Other times I stop to get a coffee as the sun is just rising and I am en route to meet people at the regional university or conduct interviews with farmers. The empanada restaurant workers offer welcoming smiles and news about their families to me. They aid me in navigating my blurred identities through simple yet profound measures of solidarity, saving me an extra-special empanada or setting aside chocolates for the children as I strive to make a holiday, like Easter, special for them though they are far from their home and their father.

As we pass a few restaurants and travel partway up from the village of Pé do Serra, meaning "the foot of the hill," we are now a half hour's travel from Ilhéus if I have been driving fast, which I always do. I feel like I know every corner of this road, which can be dangerous. We approach an overlook with a sign proudly declaring "the best view in Bahia" and park for a moment. As I stand on the cliff breathing in the warm tropical air and looking back at the now-distant city of Ilhéus, this seems to be the truest statement in the world. But I always reflect here. This is the site where, over a decade ago, Guiomar fell out of a speeding truck and narrowly survived. To this day, she bears deep scars across her back from the accident. Sometimes they peek out underneath her shirt. "I was never supposed to walk again," she often reminds me. "I lay for months in the hospital. People brought me food to keep me alive. And I knew I had to heal myself—no one else was going to do it for me. So

Figure 2 "The Best View in Bahia."

first I ate to get strong. Then I learned to sit up. Then to stand. Gradually, I taught myself to walk. At first, I could take one step. Then a few more. My goal was to walk through town to the beach." I know that this two-mile trip over cobblestoned streets with no walker must have been hell, which she confirms. "I had people who would hold my arm." She laughs at the memory and continues, "The journey would take me all day." She pauses, then continues in an upbeat tone, "But here I am!" Again, she cackles loudly as if she is victoriously laughing in the face of her close call with death. Getting back in the car, I look out the window as we pass the steep rock wall on one side and speeding cars on the other. This story is a metaphor for Guiomar's life, of near misses and survival, of a guerreira.

At the top of the big hill lies the small villa of Serra Grande, or large hill, often called simply Serra. At the first entrance to Serra, a glass structure houses a capoeira school run by two globally renowned masters of capoeira, Cabello and Tisza. The aptly named long-haired Cabello, meaning hair, is from São Paulo and has lived in New York, but over the past two decades he and Tisza have placed their bets and poured their energy into Bahia. Slowly and steadily they have constructed an impressive center for capoeira, music, and Bahian culture that is now not only a local institution but also an internationally recognized hub for capoeira Angola, a style of capoeira. They draw students from Bahia and Brazil, as well as from Israel, Spain, Germany, and the United States. I cannot pass without craning my neck to get a glimpse of people playing capoeira. I lower the window to try to catch the distant sound of the berimbau and drums that accompany the capoeiristas' movement.

I bypass Serra, which is slowly becoming a model villa comprised of nativos who live and work alongside transplants from São Paulo, other cities throughout Brazil, and the world. This recent influx of migrants has attracted people in search of peace, tranquility, nature, culture, and a sustainable lifestyle that they can construct with like-minded allies. In this modern-day alternative community, many who have chosen to come, or stay, are deeply interested in the potential for this region to serve as not only a national but also an international reference point for innovation and sustainable living. New organizations, local leadership, and development strategies are growing by the day. Money is coming in, and with it changes to the town, both good and bad, are taking place. Foundations have recently taken root here, too, in an attempt to fund innovative projects, from organic agriculture and natural foods markets to a new "slow-food" restaurant that trains youth in a profession to social entrepreneurship endeavors around art, music, and movement. A Waldorf school, now in existence for nearly two decades, is chock-full of the kids of transplants as well as locals. Things are happening here in a quiet, but vibrant, kind of way.

But there are other stories that exist alongside those of hope and deliberate, sustainable development in Serra Grande. Rumors circulate of high rates of sexual abuse, of pregnancies among very young girls, and of unspeakable things. When I visit the cemetery high on a hill at the outskirts of town, I look at the graves marked by plastic flowers. Two, I'm told, are for young boys who got mixed up in the wrong crowd, brought drugs to the villa, and were then brutally murdered. Their mother was devastated, people say. It is almost impossible to imagine the sinister tales of this bucolic place that come to me partially, in pieces of gossip, in phrases uttered as we pass a place that reminds someone of something that either happened long ago or perhaps still exists today, hidden in plain sight.

Beyond Serra, the terrain changes drastically. The straight highway parallel-
ing the coast is replaced by curves, dips, and turns. The Atlantic Forest lines the
road on both sides. On the way out of town, the Conduru State Park, Bahia's
largest and most biodiverse protected area, lies to the west. No one really goes
there save for international biologists, intermittent classes of schoolchildren,
the single park administrator who lives hours away, and the handful of guards
who occupy a small, run-down station in the middle of the park. Tourists
speed by the park's entrance on their way to the beaches; vacations in the sand,
not in the forest, are the predominant culture, for now, in Brazil.

The road continues for another thirty kilometers, hill upon rolling hill peppered with occasional houses in the distance and interspersed with cleared
pasture areas. We drive past the private Caititu Reserve, where scientists
from the New York Botanical Garden came decades ago to study and explore,
making discoveries that would make this region famous forevermore. We
pass a modest cement wall. Though it has no sign, behind it lies a private
reserve founded by a doctor who came to the region and constructed a world-
renowned research center for a very specific threatened species of snake, the
pico de jaca. For me, this wall exemplifies a regional tension. While scientists
in the center work to produce antivenom that has medicinal value, there are
other, often more common, ways of dealing with snakes here. I always think
of the day when a local friend took pride in showing us the snake he had just
killed in our garden. After beating the creature to death, he then paraded about
the streets, proudly displaying his conquest on a pole.

The world along the road continues to transform as we travel. While in
some places the forest almost engulfs the road, in others tall grasses called
capoeira lead gently toward the forest, sometimes almost camouflaging those
whose only option is to travel the highway on foot. Families carry bundles of
wood on their heads or in wheelbarrows. Mothers hold children in their arms.
Vendors of coconuts or tapioca for beachgoers push battered coolers perched
on two-wheel carts. Bikers carrying one, two, or even three passengers pass.
All move slowly, precariously, along the foot-wide shoulder that separates the
pavement highway from the capoeira and the *mata*, or forest, that lies beyond.

Over time, I come to know every pothole and curve on the road. This is
because I have chosen, for a number of reasons, to root my family somewhat
far from many of the places I work. The first is the charm of the town, with
its picturesque cobblestoned streets, mellow vibe, and staggering natural
beauty. Frankly, my family likes it here. The close second is the presence of

Guiomar, who makes our life in Bahia possible. And so, early in my time here, I resolve to do the drive from Itacaré, where we live, to Ilhéus and beyond, where I work, several times a week. Along this road, I collect my stories and those of others. Over time, our tales become indelibly woven into the tapestry of our individual and collective lives.

When I am traveling the road, Guiomar and I develop a routine to stay connected for safety. I call her if I'm leaving Ilhéus late, and we arrange a time after which, if I don't appear, she sends someone out to find me. Few people travel the road in the dark, and we both know if I break down, things can end badly. Once, they almost did. The fuscinha died on a massive hill. I rolled down backward with no power, coming to a stop at the bottom. Out of nowhere, a car appeared driving the opposite way. It stopped, turned around, and the driver offered to take me home twenty-five minutes in the direction from which he had come. However, this time I was not alone but traveling with my friend Luizinho, a tall, strong Afro-Brazilian Candomblé priest.[8] We both emphatically said no to this offer from a very strange stranger and chose to wait for the car we knew Guiomar would send when we didn't show up. We found a local bar and processed the encounter. "*Cara*, man, that was some *bad* energy back there," Luizinho exclaimed. "He wasn't from around here, and his vibe was pure evil!"

Upon our return, I did what I sometimes do when in Bahia. I lit a candle in gratitude to my deity, or Luizinho's, which were likely one and the same, and carefully placed it in the doorway to my house. Guiomar initially scoffed at my ritual. "Don't mess with that stuff," she cautioned. "You don't know what you're doing." But when I told her Luizinho had ordered me to do it to clean the house of the negative energy from our experience, she conceded quickly. "OK," she acquiesced. "If *he* ordered it, OK."

As I descend the hill that separates the municipalities of two counties, Uruçuca and Itacaré, my heart jumps not only from the steep road but also from knowing we are getting close. A large sign declares I am in the municipal district of Itacaré. With this, I know the town of the same name, our destination, is a mere twenty minutes away. Txai, one of the top luxury resorts in all of Brazil, appears to the east, discreetly marked only by a wooden sign

---

8. Candomblé is an Afro-Brazilian religion. Luizinho is a *pai de santo*, the father of the saint, a leadership position that serves as a mediator "between the material world of mortals and the spiritual world of the dieties" (Voeks 1997, 63; see also Laffitte 2010).

and neatly uniformed workers waiting at the gate to catch the bus home from their jobs. We drive past the tapioca and chocolate stand of a regional legend, a woman named Marli, and soon the road nears the town of Itacaré. Past a roundabout proudly declaring that Rotary International is active here lies the police station. Officers sleep, eat, or watch television as we pass. Sometimes we are stopped and asked for our identification or checked for having too many passengers in a car meant for five. Once, on the sage advice of Guiomar's boyfriend, Manoel, we promised to buy a fish for the officers if they would let us go. They did and we did, too, returning with the biggest fish we could find as payment for our present situation as well as an investment in any future run-ins. The next time they stopped us, we would remind them that we were the people who got them a special fish.

The road winds another six kilometers. Pousada entrances dot the roadside, hidden under arches of lush vegetation. The elite who can pay to stay here will find these places, while the rest of the population will speed by. Finally, the pavement turns to cobblestone and descends a dark hill toward the sea. Most of the streetlights are out as we careen down, highlighting the lights in the distance that mark the separation between the small town, the river leading inland, and the vast ocean beyond. To the east lies a relatively new community, named Bairro Novo, the New Neighborhood, which is largely the result of an influx of migrants to the region hoping for jobs in construction for the fast-growing ecoresorts. Amid its tranquil veneer, Bairro Novo has quickly grown to be the feared rival of the town's oldest neighborhood, Passagem, or Passageway, reputedly bringing the violence, drug trafficking, and power struggles characteristic of large urban areas to the region. I look at Guiomar and tilt my head toward Bairro Novo. "Do you still want me to stay away from there?" I kid, knowing that she gets very nervous when I venture into places she perceives as "dangerous territory." "*Deus me livre*, God help me," she replies, tsking her tongue softly and rolling her eyes.[9]

---

9. Donna Goldstein (2003) deftly portrays the complex and dynamic ways in which humor, race, violence, class, and gender come together in the everyday lives of people living in these communities as they, like the people of Bahia, "run after." Increasingly, NGOs like Catalytic Communities (in Rio de Janeiro) are leading efforts to change the public perception of *favelas*, or shantytowns, from dangerous, informal, squatter settlements to complex, vibrant communities marked by resistance, activism, and social engagement. See also Cummings (2015) and Perlman (2010).

As we come into town, the kids sit up and lean forward. They always get a bit antsy. Midway up the main street, we hear a familiar sound. A loud, cackling, high-pitched voice calling "*Oi Papai,*" or "Hey Daddy," enters our open car windows. Jeff and I look at each other and smile, knowing our next thirty minutes will be spent in a series of hand slaps, hugs, and head nods in response to questions in rapid-fire Portuguese that come one after another with no time for response. Irado, extracting his earbuds, jumps about and does a little dance, calling out to passersby while still maintaining us as the center of his attention. We get the quick report on his life. His chickens are growing well; he has built a new coop. He has three new stand-up paddle-boards, and by all means, we should meet him on the beach for a lesson. He will also drop off a surfboard tomorrow. "Just for you, no obligation," he notes with a wink. We know that with this gift will come daily visits and questions about what time we will schedule the lessons we have never agreed to.

Moving on to familial matters now that business has been covered, he produces a new iPhone and madly scrolls through pictures of his children—a pit bull mix named Pai, or father, and another named Filho, or son. "Aren't they beautiful?" he asks, and before we have a chance to respond, he says, "Why haven't you responded to my messages?" I chuckle, remembering being in the supermarket checkout line at Whole Foods or in a parent-teacher conference or hiking when a voice from another world suddenly interrupts me. "*Aí, siiiimmmmmm,*" a slang version of our "ah yeah," comes screaming through the phone as if to remind me that Bahia and our home in Boulder, Colorado, are a mere text or recording away. We extract ourselves from Irado's orbit and make our way farther down the street.

We pass one of the most well-known hotels in the town of Itacaré, aptly called "a hotel of charm." Brightly colored bungalows line the neat walkways throughout the hotel's property. There is a kids' play area close by the pool, so parents can have a cold cocktail while watching their children on ropes courses and jungle gyms. Hotel staff wear neat, ironed uniforms and bright smiles as they serve the guests. This place, in fact, is emblematic of Southern Bahia in a way. From the stunning beaches just a few steps away to the neatly arranged walkways and restaurants and tourist shops, charm is what initially draws people here.

Farther down the familiar dirt roads, we reach the home we have rented. We unload our bags, kick off our flip-flops at the entrance to keep the freshly scrubbed concrete floor clean of mud, and enter the house. Guiomar has things set, as always. There is chocolate cake, a large wooden bowl exploding with tropical fruit, and a massive arrangement of red, orange, pink, and yellow *helicônias*, bird of paradise flowers, carefully placed in the open-air living room to contrast with the dark green of the night forest just beyond. "The receipt for the supplies is here," she points out. She knows I don't ever ask for proof, but she needs to give it to me to show she is trustworthy, which I already know. We sit down and begin to truly catch up. Yes, she is still building her house, which has grown two stories since I've known her. It is always in a state of perpetual construction. Murilo, her grandson, is studying hard and developing new rap music, and yes, she would like to get him into English classes, she notes, giving me a sidelong glance to see if I'll bite at this hint, which I file away in my mind.

After this journey through landscapes and memories, through history to the present, we are back again, ready to immerse ourselves in the rhythm that is emblematic of this place. Days will begin with fishermen gathering up their nets and heading out to make an early-morning catch, which they will try to sell along the roadside just before lunchtime. A few figures walk through the streets carrying bird cages delicately high in the air. They are taking their pet songbirds "out for a walk" so the creatures can welcome the morning from their cages. Surfers will run bare-chested and barefooted along the main street to warm up before jumping in the water to catch the best breaks at dawn. While the nighttime watchmen of the pousadas traipse wearily home from their evening shifts, maids and cooks in pressed shirts will amble along the dirt road to arrive in time to make breakfast for those visiting from the South of Brazil to the South of France. Strong young men will grab a quick coffee and slice of cake before perching themselves on street corners in the hope of being picked up for day labor projects. Children will walk to school with their parents, hand in hand, as they sleepily prepare for their half day of classes.

As the sun rises, the temperatures will soar. There will be a respite in the daily hustle midday when people have lunch, nap in hammocks, or hang out in this lazy daily interval. Music may play from a car-top speaker near the *orla*, the waterfront, while children out of school for the afternoon run free on the streets or join an organized game of volleyball or soccer. Beside them on the beach, fishers will mend their nets in preparation for the next morning. A few tourists, their native guides, and the rest of the town will gather at O Xaréu, a narrow peninsula jutting between the Rio de Contas and the ocean that lies beyond to watch an impromptu capoeira demonstration as the sun sets in the background to the twang of a sole berimbau. As the sun drops, all present will gaze out at the ocean, the river, the historic Jesuit church, and the colonial buildings that line the coast where people's small fishing boats are docked.[10] Evening soccer and volleyball games will then start in the sand between the homes and the water.

Night inevitably will come, bringing a welcome relief from the sweltering heat as people rush home and the first round of the *telenovelas*, Brazilian soap operas, begin to blare from open windows. Sounds of cicadas will be truncated by the beat of congas and the call and response of songs as passersby are lured to view the capoeira *roda*, the circle that encompasses the white-clothed capoeiristas. Tourists will stroll the cobblestoned streets, looking into shops that only open in the late afternoon and evening, for those who travel here are at the beach during the day rather than shopping, of course. Local restaurant staff will stand on the street, trying to attract customers by announcing their culinary delights. Occasionally, a sleek SUV will slowly roll by, perhaps carrying a supermodel or a European president. "They all come here," nativos proudly claim while some, like Guiomar's son Neto, sometimes venture off to try to make a life la fora. As a friend of mine from the region likes to say, "Lots of people come here, stay for a while, and then go. But *we* must find a way to make our lives here."

---

10. The Rio de Contas, the state's largest river basin that affects eighty-six different municipalities, starts in Western Bahia and empties into the Atlantic Ocean in the town of Itacaré, which was founded by Indigenous tribes (including the Tupiniquim, among others) and then, in 1532, colonized by Portuguese. See Chiapetti (2009) on the significance of the Rio de Contas to residents of Itacaré. São Miguel church, built in 1723 by Jesuits priests who settled in the region, is a landmark in Itacaré.

A Very Brief Orienting Interlude

Cultures, Politics, and Interconnections across Difference

At a cursory glance, the Southern Bahia region of Brazil's Atlantic Forest hardly seems like a place of conflict. Far from the violence of large northeast port cities like Salvador or Recife or the frenzied chaos of Brazil's megacities, the natural environment of endless pristine beaches lined by verdant tropical forest provides a stunning backdrop for an ostensibly relaxed, carefree lifestyle. Yet the tranquil veneer of this region masks more nuanced realities. While diverse social movements and activist groups have been present here for decades, tensions and ruptures can surface, particularly between and among people when they address issues of forest conservation, livelihoods, land rights, and development. People struggle to collectively define what it means to be *of* and *in* this place, which can unearth both historical and present-day inequities and differences.

At times, divisions among those living here can be pronounced. Environmental activists can assert that land reform movements or independent family farmers pose threats to the region's globally important conservation agenda. Farmers can resent natural areas being reserved for plants rather than people and complain that what they view as an elite environmental movement is disconnected from the plight of the poor. In similar yet distinctive voices, *quilombolas* can mourn the loss of land and livelihood opportunities to what they perceive as esoteric and historically blind conservation ideals, while nativos can resist the ways in which capital, power, and unchecked neoliberal development strategies are being imposed on the lands they have long inhabited.[1] Activist leaders of these diverse social movements sometimes bemoan their inability to work together with a simple, yet telling, phrase: "They don't speak our language." Today, though some of these long-standing rifts are dissipating, to be sure, the future of Southern Bahia remains uncertain.

In this book, I construct a history of the present in the Atlantic Forest of Southern Bahia; that is, I draw upon the region's contested past to reveal its current struggles as well as its future possibilities.[2] In this journey, I seek out and strive to make sense of what anthropologist Anna Tsing aptly calls "politics-in-friction," or heterogeneous, unequal, and contingent encounters across difference.[3] When emotions rise up or outbursts take place, when people make statements or do things that don't seem to align with "who they are," I pay attention. The meanings and materiality of these inconsistencies and incompatibilities matter, for they indicate a deeper cultural politics at play.[4]

1. Quilombolas, or descendants of those who resisted and fled slavery and established independent maroon communities, called *quilombos*, will be discussed extensively in chapter 3. Neoliberal development, which will be discussed in chapter 4, is a capitalist, market-based approach that privileges privatization and commodification of natural resources. See Ong (2006) and Harvey (2007) for foundational texts on neoliberalism.

2. In using the term "history of the present," I draw from French philosopher Michel Foucault's (1977) emphasis on power and political struggle; we need to understand history as a process for revealing the contingency of the present, which can in turn be instrumental in creating a more open and inclusive future that works to rectify historical inequities and power imbalances. See Garland (2014) for an interpretation of this concept.

3. Tsing (2011, 14). Anthropologist Paul Rabinow and colleagues assert that some people are drawn to searching out "what is going on below, beyond, and between acceptable discourse" (and, I would add, practice) (2008, 23). I am one of these people. Rabinow et al. observe that this type of critical analysis and thinking leads to future possibilities, or "operating in the future inter-zones" (23). I am drawn to how conflict can be productive by helping us to recognize past injustices and, in turn, co-create a more inclusive and equitable future.

4. Throughout this book I engage regularly with the ideas of the American-Colombian anthropologist Arturo Escobar, who writes extensively on the relationships among culture, biodiversity conservation, and social movements. Escobar describes cultural politics as "the process[es] enacted when sets of actors shaped by, and embodying, different cultural meanings and practices come into conflict with each other. . . . Culture is political because meanings are constitutive of processes that, implicitly or explicitly, seek to redefine social power" (1998, 64). Through the stories to follow, we will see how culture travels through individuals, as well as through the broader social movements of which they are a part, to both perpetuate as well as reshape and reconstitute power relations.

Histories of colonialism and clientelism, present-day constraints of race and class, and myriad forms of "Othering" can create incongruities, misunderstandings, and fissures among contemporary leaders, social movements, and activist groups in Southern Bahia.[5] Recognizing, discerning, and addressing conflict and difference can make visible the underlying power relations, which in turn can shape how people view themselves as well as how they perceive of "the other." Again, this matters, for, as anthropologist Fernando Coronil observes, "in the context of equal relations, difference would not be cast as Otherness."[6] And when otherness and difference are firmly rooted in historical and/or contemporary power inequities not only can it be difficult to "speak the same language," sometimes it can be hard to speak (or be heard) at all.

This, too, matters because social movements arise for specific reasons. They are built from collective identities. They emerge out of common histories of injustice and demands for systemic change. In the complex, contingent, and intersectional relations that shape social movements in regions like Southern Bahia, we find struggles, striving, and efforts to move ahead amid all of the contingent complexity that this involves.[7] People's intentional choices and actions around these struggles have important ramifications on landscapes, on the governance of resources, on their ways of making their lives in the places they live.

These forces at play in Southern Bahia today also raise salient and timely questions. Will people in this region adopt and adhere to environmental conservation rules and regimes, or will they be left with few choices but to harvest the paucity of remaining Atlantic Forest for timber, hunt endangered animals, and engage in other extractive livelihoods, which undermine the sustainability of the region's forests and waterways and its ability to maintain biodiversity and ecological health in the long term? Will people successfully confront and combat the stark realities and challenges of this place, or will

---

5. Johannes Fabian describes "Othering" as a process that takes place through interconnection across difference: "Othering expresses the insight that the Other is never simply given, never just found or encountered, but made" (1990, 755). See also Said (1978).

6. Coronil (1996, 56).

7. At the core of evolving relations across social movements lies what anthropologists Donald S. Moore, Anand Pandian, and Jake Kosek call the "*simultaneity* of symbolic and material struggles" (2003, 40, author's emphasis), struggles over culture and meanings that also involve struggles over socioeconomic conditions (Escobar 1992, 412).

they allow rampant poverty and limited livelihood opportunities, rural-to-urban migration, and precarious education, healthcare, and infrastructure to perpetuate? Will new partnerships and alliances among historically disparate groups arise, or will long-standing constraints of opportunity and equal access preclude new possibilities for collective action? Will people here be able to successfully arrest neoliberal development schemes imposed on them by external actors, or will they instead find ways to ally, imagine, and establish durable and diverse livelihoods that preserve the region's cultural and environmental characteristics? Questions like these are simultaneously contextual and universal, timely and timeless. They are embedded within the cultural politics of a place like Bahia, but their answers have application well beyond. And uncovering the complex and contingent answers to questions such as these allows us to better understand complex and dynamic landscapes like Southern Bahia, how people are of and in a place. They reveal the individual stories and collective story of here.[8]

---

8. New frameworks of analysis are crucial for advancing social movement studies both within anthropology as well as across disciplines. For example, see Escobar's (2008) compelling and creative ethnography on how history, geology, biology, activists' practices, capital accumulation, development and conservation strategies, and cultural politics shape Afro-Colombian social movements.

Narrating Environmentalism in a Biodiversity Hotspot

"We're seeing now that Southern Bahia is living in a critical moment. If we don't have an agenda of development compatible with conservation that sustains the conservation of the forests, we're going to have thousands of problems in the future."

—Rui, Ilhéus, Bahia

Creating Place

I drive along the roads of Southern Bahia engaging in a mental exercise by counting all references to "eco" in any form that pass before me. The word is laden, like the forest itself, with context and innuendo depending on who you are or where, or how, you work. Along the highway that traverses the coast, nailed to a tree is a hand-painted board advertising "eco-wood" for purchase. I pass several brightly colored welcoming signs for ecoresorts. There are no formal certification programs or standards in place here, yet the majority of tourism lodges have some version of eco associated with them. At the entrance to one of the region's towns is an advertisement for eco-cement, which can be used to build your ecoconstruction, the sign claims. I begin to lose count in my little game. It is a dizzying narrative of all things eco. Ecotrip, ecosafari, ecochic, ecoluxury, ecohotel, ecocuisine, ecovilla, ecococktail. Movements, like the environmental movement, are fueled by narratives—histories, stories, words, even practices that convey purpose and meaning. But while this environmental rhetoric is seemingly ubiquitous in the repeated use of the term "eco," it condenses two realities of environmentalism in this region into overly simplistic portrayals.

One reality, the region's staggering biodiversity, is indisputable. The ecological importance of Southern Bahia's Atlantic Forest is counted, registered, verified by scientific studies, transects, records, and recordings of flora and fauna. But the other reality of the Atlantic Forest, how certain people feel about the environmental movement here, can be more obscure. This cannot be simply documented, for it is less visible and overt; it is shifting and sometimes hushed. The importance of and the relationship between these realities, however, cannot be denied.

The Atlantic Forest has long been famous throughout the global conservation community for dramatic statistics grounded in biological facts. As Warren Dean, the most noted historian of this biome, describes:

> On the eastern margin of South America there once stretched an immense forest, or more accurately, a complex of forest types, generally broad-leaved, rain loving, and tropical to subtropical, stretching from about 8° to about 28° south latitude and extending inland from the coast about 100 kilometers in the north, widening to more than 500 kilometers in the south. Altogether the forest covered about a million square kilometers. This complex has been referred to as the Brazilian Atlantic Forest, related to the much larger Amazon Forest but distinct from it. Together, these two great forests formed a life zone distinct from and richer in species than those of the other tropics of our globe, situated in Africa and Southeast Asia.[1]

Today, just over 12 percent of the original Atlantic Forest remains standing, and the Southern Bahia region has less than 5 percent of its original primary forest.[2] The remaining Atlantic Forest is designated a UNESCO biosphere reserve as well as a global "biodiversity hotspot." Biologists call it the Hileia Baiana, a name coined more than five decades ago that marks the region's incredible levels of biodiversity and endemic species, or plants

---

1. Dean (1997, 6). This historical account, one of the most comprehensive records of the entire Atlantic Forest to date, traces the human and environmental interactions in this region over the course of its long history.

2. SOS Mata Atlântica (2019) reports 12.4 percent of remaining original forest is covered under the Law of the Atlantic Forest (*a Lei da Mata Atlântica*). For past studies on the importance of the Atlantic Forest region, see Ribeiro et al. (2011), Myers et al. (2000), Tabarelli et al. (2005), and Leal and Gusmão Câmara (2003).

Figure 3 The Atlantic Forest.

and animals found nowhere else.[3] In the scientific community, there is also another term that communicates the uniqueness of Southern Bahia's Atlantic Forest: it is dramatically called "a hot-point within a hot-spot."[4] All of

---

3. Faria et al., who bring together Southern Bahia's history, ecological importance, and conservation potential in a deft interdisciplinary analysis, note, "The Hileia Baiana shelters one of the five centers of endemism along the Atlantic Forest biome harboring one of the most diverse areas for plants and animals in the world" (2021, 64). Thomas, Garrison, and Arbela (1998) found that 41.6 percent of the plant species collected in Serra Grande are endemic to the Atlantic Forest along Brazil's coast and 26.5 percent are endemic to Southern Bahia and Northern Espírito Santo (a state just south of Bahia).

4. Martini et al. (2007). The term "hotspot" was first used by British ecologist Norman Myers in 1988 to describe highly threatened areas with large numbers of endemic species, meaning plants or animals found nowhere else on earth. In 2000, Myers et al. identified twenty-five global hotspots that together comprise only 1.4 percent of the Earth's surface yet contain 44 percent of all species of higher plants and 35 percent of all land vertebrate species, and characterized Brazil's Atlantic Forest as one of the five most important global hotspots for floral biodiversity (2000, 853; see also Brandon et al. 2005).

these names, terms, and descriptions convey an important message: this is one of the most threatened ecosystems on Earth.[5]

Almost daily, I travel along the regional highway that runs north–south between the coast and the forest. Without fail, halfway between the towns of Itacaré and Ilhéus, my eyes dart to the east as I look for a particular plaque on the side of the road. For most who pass, this is nothing other than a sign denoting a private reserve. If you drive too fast, you can easily miss it. But this sign marks a place that has forever changed the way the world perceives and characterizes Southern Bahia's Atlantic Forest. In the late 1990s, on the Caititu Forest Reserve, biologists from the New York Botanical Garden recorded a world record number of tree species in a single hectare. *This* is the discovery that placed Southern Bahia on the ecological map, so to speak, revealing that parts of this forest have more floral and faunal diversity than the globally famous Amazon.[6]

But while UNESCO Biosphere Reserve, hotspot, and hotpoint designations initially drew me to this region, there is more beneath the surface that sometimes counters the veneer of "all things eco." This narrative is less public. It is expressed as people mutter under their breath or roll their eyes when environmentalism is discussed or reduced to the "eco" references that are so pervasive they are sometimes meaningless. One hot afternoon, I sit with a few men who live near Southern Bahia's largest and most biodiverse state park. We discuss soccer and the upcoming Carnaval celebration, and soon our conversation shifts to the region's environmental identity. We talk about the beach, the forest, the monkeys that can sometimes be spotted

---

5. In 1992, UNESCO designated much of the Atlantic Forest a World Biosphere Reserve, meaning a globally important reserve intended to unite conservation and sustainable use. In 2002, the rest of the Brazilian Atlantic Forest was awarded this status. Such designations are not unique to Bahia or Brazil. Furthermore, they are complex and contested spaces. Runk (2017) explores how particular landscapes, like Panama's Darién, are complex, dynamic (and constructed) conglomerations of biophysical characteristics, politics, and cultural beliefs. Martínez-Reyes (2016) looks to how colonial inequalities are reproduced through ethnic relations around biosphere reserves in Mexico. West (2006, 4) articulates how conservation spaces that are in "ostensibly out-of-the-way places exist within, are made by, and help to make transnational loops." All of these concepts apply to Southern Bahia.

6. Morellato and Haddad (2000, 786). See Piotto et al. (2009) for an important study on how tree species in secondary forests in Bahia take more than forty years to recover the structure of old-growth forests as secondary forests show species richness but not similar forest structure.

crossing the main highway here on fragile rope ladders strung high above the trees. Suddenly, one of the men stops talking, looks at me squarely, and flatly declares, "You know, environmentalists are turning people from here into robbers—they are *making* us rob. When you tell a guy who is hungry that he can't cut a tree or hunt to feed his family, you are making him turn to crime to do so." His friend looks on and nods, offering his own interpretation of environmentalism in a biodiversity hotspot: "Sometimes it feels like environmentalists are taking the ground out from underneath those who are already living on the ground!"[7] He chuckles at his clever reference, takes a long swig of Coke, and looks out at the forest that lies just across the roadside where we sit and nods. I am slightly disturbed. Some of my best friends are environmentalists. *I* consider myself an environmentalist. And I have come to this region to study environmentalism.

Connecting and *Confiança*, or Trust

The environmental community in Brazil is a relatively small, tight-knit group of people who have often worked closely together, or at the very least crossed paths. This is despite the fact that Brazil is a massive country (the fifth largest in the world) with six major biomes, or distinctive areas of vegetation, wildlife, climate, and soil.[8] I landed in Southern Bahia through contacts and connections that began not in the region itself but through people who knew this place, who introduced me to the importance of environmental activism here. It all began years ago on a small sailboat in Salvador, the capital of the state of Bahia.

Mario and Paula are on a mission. I'm along for the ride. We wander through Salvador's upper city, passing through a cemetery and heading eastward

---

7. DeVore (2018) found a similar tension between squatter farmers and environmentalists in Southern Bahia. See also Zanotti and Knowles (2020) on how political ecology's attention to power and identities (among other things) around intact forested landscapes illuminates the need to consider multistakeholder approaches and local actors' knowledge and needs to better address poverty alleviation related to forest governance and landscape management. Vivanco (2006) also portrays the complex dynamics and "culture of nature" that exists around Costa Rica's Monte Verde forested region.

8. Brazil's biomes are the Amazon Forest, Atlantic Forest, Cerrado, Caatinga, Pantanal, Pampa, and Coastal.

toward the water. The moonlight shines across the graves of varying heights that stick up from the ground, arranged in neat rows like antique Legos adorned with faces, names, statues, and sometimes plastic flowers. These surroundings, which would normally be eerie if I were alone, are peaceful with good friends. A few kids in dirty, tattered clothing ask us for change while their friends sneer, sniff glue, and then disappear into the shadows behind an angel adorning one of the graves.

Descending an elevator that travels from the upper level to the lower part of the city, along the water, Mario and Paula move increasingly quickly. Their unwavering determination, despite the enticing distractions of the urban streets, indicates that pausing is not an option. We navigate our way through Baianas, hefty Afro-Brazilian women wearing starched white dresses with lace borders and white turbans who sell hot, spicy *arajé*, or beancakes, to hungry passersby. Vendors peddle small replicas of churches, saints, and even miniature Baianas that are neatly arranged in rows to attract the multitudes of strolling tourists. A few young guys in white pants and bare chests play a berimbau. They hurl into the air, doing twists, turns, and somersaults as they do capoeira. Slavery in Brazil began more than five centuries ago just steps from where we are, near Salvador's port, where ships arrived from Africa. But I cannot stop to ponder this for long.

Instead, I trail closely behind Mario and Paula, observing and listening. This is what I have learned to do. The three of us have developed a habit of meeting up in airports or for frenetic dinners at environmental conferences, moments that challenge us to cram years of conversation into our fleeting time together. Paula and I met more than a decade ago at an ecotourism conference in Washington, D.C., and then reconnected deep in Brazil's Amazon in the city of Manaus. Our instant friendship was fueled by long nights discussing development, environmental conservation, men, and life. We would chortle deep belly laughs as we shared our stories. Now, Paula is my steadfast go-to in São Paulo, the person who picks me up wherever I am, who finds time to take me out for a good meal or to gather in the home of our common friends, many of whom are leading academics and activists in Brazil's environmental movement. When she sees me, she invariably shouts, "*Eba*," meaning yay. And after our time together, from minutes to days, upon departure she will open the trunk of her car and produce an interesting gift. Artifacts and artisan work from her trunk dot my home, a delicate handmade flower here, tiny espresso cups painted with the Brazilian flag there. While these things physically link her world and mine, our personal and intellectual connections are equally strong.

Mario, too, is a frenetic bundle of energy and intention. He is a one-man institution who spreads infectious energy throughout the Brazilian, and global, environmental community. One day he's in Brazil's capital, Brasília, lobbying lawmakers on technicalities of the country's Forest Code.[9] The next he's in his hometown of São Paulo, where he works for one of Brazil's largest environmental organizations, SOS Mata Atlântica. Another day he's in Paris at the United Nations Conference of the Parties. He appears on television during the nightly soap opera slots doing information briefs on forest restoration; he's in the far-off Amazonian town of Porto Velho meeting with state environmental officials about a new law. He is ubiquitous. And each time we meet, our encounter is marked by a booming greeting, some jumping up and down in a Brazilian style of a jig, an ample smile, a loud laugh, a bear hug and kiss with his scratchy beard, and then, depending on how much time we have, minutes to hours of joyful revelry interspersed with serious, thoughtful conversation about the state of environmentalism in Brazil.

But on this night I am not thinking of this history among us. Instead, we are merely moving quickly. They are focused on a destination, and I am tagging along. We cross a busy coastal road and head toward the darkness of Salvador's harbor, which is punctuated only by a few twinkling lights on the moored boats and the soft reflection of the city's glow far off in the distance. We walk along the dock for several minutes. Mario and Paula are on a search for something. Soft, rhythmic waves gently slap the dock, truncated by our loud footsteps on the metal planks. They pause, then slowly continue on, staring intently at every boat we pass. Finally, one of them exclaims, "This is it!" Mario climbs aboard a small sailboat, and Paula and I follow suit. A million stars carpet the sky above us while we stretch out on the boat deck. Three people, on a boat, in a harbor, off the Atlantic coast of Brazil. The only

---

9. Brazil's Forest Code is national legislation that was first passed in 1965 and has been subsequently revised, with the latest revision in 2012. Analysis on the relationship between the Forest Code, Brazil's greenhouse gas emissions reduction targets, and the potential as well as challenges of Forest Code implementation include Azevedo et al. (2017), Soterroni et al. (2018), and Soares-Filho et al. (2014). As will be drawn out in other examples throughout this book, as in land designation and tenure, Brazil often suffers from progressive laws on paper with spotty implementation in practice.

sound is the ever-so-faint music of street parties in Salvador echoing in the distant background. Mario and Paula are silent, as am I, as we gaze at the sky.

"Damn, he was an amazing guy," Mario says quietly. "What a great environmentalist, what a leader." "Yeah, he was," Paula agrees. They are silent again. Almost as an afterthought, Mario turns to me as if he suddenly remembers that I don't know why we are here or who *he* is. "This is the boat of our friend Pedro. He loved to dive. He was an important leader in the environmental community throughout Bahia. He really fought for this region, creating more than twenty Environmental Protection Areas across the state," Mario starts. He pauses as his voice catches. Paula takes over where he leaves off. "But one day Pedro dove too deep. It must have been about a year ago. He never came up. . . ." Her voice trails off, leaving the obvious unsaid. "He left behind a wife and small child," one of them says. "They were waiting for him on the boat."

We drift back into a collective silence, holding the space together in a way that seems sacred. As Mario and Paula remember their friend Pedro, I contemplate Max, my nine-month-old son at home. I know that I, too, am the type of person prone to underestimating how much I will need, not in diving but in life. I know deep inside my bones that I, too, could very well be the person to miscalculate and never surface.

But I also know that the community of those around us is what connects us, what brings us to the surface when we are pulled into other experiences. My bond with Mario and Paula continues to this day. They are the ones I turn to for expertise in all things environmental in Brazil, and they are the ones who initially inspired me to work in Bahia. Perhaps this visit to Pedro's boat was subconsciously intended to bring me into the orbit of their friend who had perished in the sea but whom I still needed to somehow know.

Years after our time together in Salvador, I head to Mario's office at the SOS Mata Atlântica headquarters. It is a chilly July evening in São Paulo. I have left my family and traveled from the altiplano of Brasília to the Pontal do Paranapanema in the interior of São Paulo and now to downtown São Paulo. All of this is in search of a good field site. I have spent eighteen hours on a bus marked by intermittent stops for food along the way and maddeningly cold air-conditioning. I have conducted interminable interviews with anyone who would talk with me in my effort to find a place where I can study the complex intersections of environmental conservation and farming in Brazil. Now, I am glad to be staying in Paula's home in São Paulo and to have a chance to visit Mario in his office to report on my travels thus far. It is late in the evening, and most of the staff are heading home, but I know Mario will still be there and now, with the passing of the day, he will be more relaxed and have more time to chat. He will certainly have an opinion on where I have been, on what I have seen, and on where I should go next.

Mario and I lounge and fuel ourselves with a little evening coffee, or *cafezinho*, and start to trade stories about the Atlantic Forest. He shares decades of insight and expertise on his work. His opinion is invaluable to me: he has encyclopedic knowledge to share with all who will listen, and I am one of the rapt and willing. Upon learning that I am headed north to Bahia next in my extensive field site quest, he enthusiastically exclaims, "Well, you need to meet Rui! Let me call him now!" I stammer awkwardly, "No, you don't have to, really." I am still honing my Portuguese language skills and am uncomfortable with the Brazilian norm of just calling people and expecting them to hop on the phone instantly. Mario ignores my protest. In Brazil, relationship building and maintenance are like a finely tuned orchestra, and he is a seasoned maestro. He picks up the phone and places the call, oblivious to the late hour and my obvious apprehension. "Rui!" he exclaims excitedly in his megaphone-like voice. "My friend is coming to you tomorrow. You need to meet her."

Meeting Rui

When I first meet Rui, he is starting a new environmental NGO in Southern Bahia called Instituto Floresta Viva, meaning the Living Forest Institute. I am nervous, as I always am with new beginnings. I park the fuscinha outside the Floresta Viva office, which is located in a simple space at the top of the (in)famous "best view in Bahia" hill in Serra—the hill that almost claimed Guiomar's life. The office building is sizable, marked by the logo of the region's rural agricultural producers' association neatly painted on the bright yellow exterior wall. I peek inside to see a wide-open space with a simple concrete floor and benches lining the room's periphery. It seems anything can happen here—from storing agricultural produce to hosting a massive dance party.

Gradually, farmers in rubber work boots start to gather outside, greeting each other with handshakes and claps on the back. I wait. Over time, they slowly file inside. I wait. A few women enter, some dragging young children by the hand, and largely keep to themselves as they take seats along the benches. I wait. People murmur in low tones. It gets hotter as the morning draws to a close. Still I wait, all too aware that I stand out as someone *de fora*, the person from outside, but unsure of how to bridge this obvious gap.

I intuitively know Rui as soon as he enters. He is small in stature but large in presence, with a sharp jawline, piercing eyes, and a mane of dark, wavy hair. Going around the room to greet all who are present, he shakes hands and pauses, focusing on each person. Some he embraces with a combination of a handshake, hug, and back pat. Finally, he arrives at me, gazing intently as if trying to place me in this group. I stammer an introduction as he patiently listens to my accented Portuguese. Suddenly, his face lights up with a gentle smile in recognition of Mario's phone introduction. "Welcome to Southern Bahia," he states simply. "We're glad you're here."

After the greetings, Rui takes his place in the front of the room. He begins by conducting a visioning exercise with a group of farmers involved in a conservation program that Floresta Viva is running. He draws two big circles on the floor, slightly overlapping, labeling one forest and the other agriculture. Shading in the middle, he looks up as people gather around to see what he is doing. "Here is where we are." He points to the shaded area. "How do we promote sustainable agriculture that helps people and preserves the forest?" As people start to chime in with their responses, some get up and stand behind him to contribute their ideas. For the next hour, Rui crouches patiently, listening and drawing a collective vision for the region as it rises up from comments, offerings, ideas of those surrounding him. I leave this participatory exercise feeling that he has a knack for drawing people in, for listening and co-creating something, though I am not yet sure what this something is.

Recreating History

To understand the Atlantic Forest in the present, we must consider its past, a story that began centuries ago. On a hot day in February, a small group gathers in an air-conditioned room at one of the region's most prominent ecoresorts. The nativo guides, youth from the region, are eager to develop their options for working in the fast-growing ecotourism industry. Many of them are used to interacting with the public to some degree already—they are local capoeiristas who perform for the tourists who stroll the streets of Itacaré in the evening or they work as "bar men," standing on the street curbs adorned in surf shorts and an apron as they make fresh drinks for passersby. But this afternoon, the group shifts in their hard wooden chairs as they struggle to stay alert in a sterile, windowless room while the ocean waves pound audibly just beyond.

Rui stands at the front next to a large whiteboard. "Who can tell me about the history of this region?" he asks. The room is silent. Some of those present take sidelong glances at each other to see who will bravely speak first. Rui waits a moment and, with still no response, continues. He is intent on capturing the 500-year history of Southern Bahia in a short, interpretive narrative that will engage the group. He understands that those present not only need to know this story but also need to be able to communicate it. And so, he launches in with a brief, informal but descriptive interpretation of Bahia's history of colonization.[10] "The colonizers here came from Europe," he begins. "The Portuguese arrived here, found the Indians, had children with them, and also displaced them from their land. This started the process of taking products away. We call this the period of *pau brasil*, Brazilwood. Demand was created. People thought, 'If I arrive in Portugal with a full load of pau brasil, I'll be so rich, and I, as well as my children and grandchildren, will be able to live for the rest of our lives without working.'" Some of the youth nod, following this complex connection among colonization, miscegenation, and natural resource exploitation.[11]

---

10. Colonial forces shaped the coastal Atlantic Forest not only through rampant timber extraction but also through how land was conceived, divided, and demarcated. The Portuguese divided the country into *capitanias*, or capitancies, which condensed a vast geographical area larger than the continental United States into thirteen territories running from east to west, with each possessing a valuable (though varying in size) tract of coastline. The primary gateway to Southern Bahia, the city of Ilhéus, was one of the original capitancies, called the Capitania de São Jorge dos Ilhéus. These capitancies were distributed by King Dom João III to prominent Portuguese families, creating a system in which the elite few politically, geographically, economically, socially, and even racially dominated the entire country (see Page 1996; Skidmore 2009). This system gave way to landownership systems and inequities in Brazil that continue to this day. "The *latifúndio*, or large estates, originated at the birth of the colony and remained a dominant characteristic of Brazil thereafter" (Burns 1993, 27).

11. Rui's description aligns with that of Warren Dean, eminent historian of the Atlantic Forest, who interprets, "One of the first acts of the Portuguese mariners who reached, on 22 April 1500, the forest-overladen shore of the South American continent at 17 degrees south

I think back to an afternoon a few weeks ago, when Guiomar's boyfriend had reconstructed this same colonial history of the Bahian coast with my children using sticks on the beach. He had portrayed colonization along this very coast, from Portuguese ships and cannons to the invasion of Indigenous peoples' communities. I also think about the complexities of this dynamic environment, from the extraction of timber and the original cacao exports from Brazil to Europe that fueled the colonial project to the present-day possibilities labeled as "eco" that perhaps could support a regional economy in dire need of economic development and sustainable livelihood options. I also think about history and the present colliding in this very room, about the ways in which intersectional dynamics of gender and race and class have resulted in this group of almost all male Afro-Brazilians of very modest economic means that is gathered here today in search of opportunity. These entrepreneurial youth are running away from the deeply rooted histories of oppression, from the denial of opportunities for them to truly be *of* and *in* this landscape. They are running after, running toward, new ways of belonging. The question at hand is if and how they can forge these new pathways in racialized landscapes and if the region's long-standing and deeply engraved inequities of power are imprinted indelibly or, instead, can be rewritten.

Rui's words call me back to attention as he continues connecting the region's past to its present. He looks at the group and says, "People started to plant a little cacao,[12] and the region began to grow with cacao, while in

---

latitude, was to cut down a tree." The colonial invasion initiated a new approach to the Atlantic Forest region, one based on massive timber extraction rather than agriculture or hunting, which was the tradition of the Indigenous Tupí and Guaraní tribes that had been living here long before Brazil's colonization. Dean reveals how human occupation of the forest, though poorly documented, involved Indigenous Tupí and Guaraní engaging in 10,000 years of hunting and gathering, followed by a thousand years of swidden farming, *before* Pedro de Cabral landed on the southern coast of Bahia to colonize Brazil in 1500 (Dean 1997, 65).

The literal and symbolic tree felling with colonization sparked a trajectory of exploration, exploitation, and demarcation, forces that continue to exert influence on Bahia to this day. Relatively quickly, the coast of Bahia became known as one of a few places in all of Brazil to extract the tree that named Brazil, *ibirapitanga* in Tupí, or *pau brasil* in Portuguese, due to its bright orange-red color when cut, which was like the word *brasa*, or coal, in Portuguese. The presence of pau brasil marked the beginning of the discovery—of naming, claiming, and place-making that has characterized the Atlantic Forest in myriad ways ever since.

12. Cacao (*Theobroma cacao* L., also called cocoa) originated in the Amazon region of Brazil and was introduced into the town of Canavieras in Southern Bahia in 1746 (Mori and Silva 1979; Piasentin and Saito 2012).

Salvador, many people were working with sugarcane.[13] All of the coast sent products in the direction of Salvador." He points north, indicating the route of export. "Itabuna, the region's largest city," Rui continues, "barely existed in 1910. But in the twentieth century, the plantation population exploded! People came here to make large cacao plantations because everyone in Europe in the nineteenth century had begun to consume these products."[14] He engages his audience with rhetorical questions. "These products came from where?" He looks around as the youth shuffle in their seats and responds to his own question in professorial style. "From agricultural places!" He concludes the lesson with a final question. "And cacao makes what? It makes chocolate!" and reiterates, "Chocolate became popular in Europe in the nineteenth century. Everyone in Paris and Europe had to have chocolate."

Rui pauses his history lesson and looks around at the faces gazing back at him. Many are descendants of formerly enslaved people, who planted sugarcane and cacao, feeding the colonial system.[15] They are all hoping to break into the ecotourism industry that depends on the verdant Atlantic Forest surrounding our small room in this ecoresort on one of the most beautiful, and private, beaches in the region. It seems as if Rui realizes he needs to make this history even more real to them, so he underscores his rationale. "I'm telling you all this so you have an idea of what the Southern Bahian region is today. So you understand *why* this region has forest." He picks up a chocolate bar and holds it high in front of the group. "People across the ocean wanted *this*," he says, shaking the bar for emphasis. "And *this* is why we still have forest here!" he exclaims. "Because cacao grows best underneath the forest

---

13. The Atlantic Forest regions to the north of Salvador were largely transformed by the sugarcane industry (Rogers 2010; Scheper-Hughes 1992), while the Atlantic Forest in southern Brazil was indelibly marked by rubber and coffee production (Dean 1997).

14. Caldas and Perz note (2013, 149), "The development of the plantation economy in southern Bahia encouraged further migration and population growth, and the profits generated by cocoa production permitted regional development and integration of southern Bahia with the rest of the state."

15. Intersectionality, or the myriad ways in which race, class, and gender come together to discriminate and privilege, is evident throughout Southern Bahia as a remaining legacy from colonialism. Light-skinned descendants of European settlers, for example, were often the property-holding elite, while Afro-Brazilians bear histories of slavery, as will be discussed in chapter 3. These historical dynamics create racialized and gendered landscapes of belonging that are still negotiated in the present. See Perry (2016) for a discussion of intersectionality in Salvador, Bahia's capital.

canopy." He pauses, then dramatically concludes with the truth: "And this region *breathed* cacao."

Though cacao originated in the Amazon, Southern Bahia is known as the cacao region of Brazil. Like all things eco, this is also a place where, for well over a century now, Southern Bahia has also been all things cacao.[16] While this history is visible in the countryside today, cacao also marks the region's potential for the future. It is omnipresent. Chocolate can be purchased the moment you land in the airport. T-shirts bear large yellow cacao pods. We almost always have fresh cacao in our home; when my children come home

---

16. Between 1818 and 1872, the population in Southern Bahia more than doubled and then continued to soar over the next three decades, going from 2,400 residents in 1818 to 10,000 in 1881 and to 105,259 by 1920, writes Mary Ann Mahony (2016, 92), a historian of Southern Bahia. By 1890, Bahia was one of the world's most important cacao producers. Mahony describes how cacao, settlement, and race are intertwined, painting a vivid picture of the cacao rush in the region. Describing early Afro-Brazilian cacao farmers, she writes, "These people had acquired their land well before abolition and, more importantly, before the prospect of growing cacao drew thousands of migrants to the Ilhéus frontier. Most small farmers, Afro-Brazilians included, obtained their land later, in the Ilhéus land rush of the 1880s and 1890s." Quoting from Bahian novelist Adonias Filho, Mahony recounts how farmers would set off past the existing plantations up the rivers and over the hills into Ilhéus's far back country: "they launched a veritable invasion of the southeastern Bahian forests. Most of them squatted on public lands located in the western and northwestern districts of Ilheus. . . . By 1905, most of the land beyond the coastal sesmarias [a land designation term] but within about two weeks travel of the coast had been claimed and at least partially planted in cacao. In the next ten years, rapid planting continued, and by 1920 there were over 6,600 cacao plantations where forty years previously there had been only virgin forest and Botocudo Indians. In 1897, a land law, passed by the state legislature under the aegis of Brazil's 1850 land law, made it possible for these squatters to obtain legal title to the land that they claimed" (Mahony 2016, 98). See also Walker (2007) on slave labor, which produced a vast majority of cacao until the end of the nineteenth century.

The region's most famous Brazilian author, Jorge Amado, has penned tomes about the cacao era in Southern Bahia, a time when those most connected to the colonially established structures of power were the ones who profited most from cacao production, called "white gold" due to the color of the cacao pulp and pods. Amado describes the realities of the region in his book *The Golden Harvest* (1992). Ilhéus is "a city of money and cabarets, of dauntless courage and dirty deals." On the cacao plantations, "the cacao fields are the work, the home, the garden, the cinema, often the cemetery of the workers. The enormous feet of the hired hands look like roots, bearing no resemblance to anything else. The gummy cacao sticks to their feet and never comes off, making them like the bark of the trunk, while malaria gives them the yellow color of nearly ripe pods, ready for picking" (1992, 94). See also Lamberti (2017) for a comprehensive analysis of cacao's role in the economy, history, and culture of the region.

from school, they clamor to open the pods and suck the sweet, milky seeds inside. Many days, we make thick, delicious cacao juice. Our favorite grocery store chocolate is named Ouro Branco, or white gold, after the soupy mucus that surrounds the beans. Cacao also bears its mark in the region's social relations, in land distribution, in memories of the past, and in hopes for the future.

Understanding Cacao

Driving on the narrow, curved roads of Southern Bahia's countryside, or interior as it is called, I work hard to avoid swerving into traffic, pedestrians, and bikers as my eyes are lured to the cacao farms. These lands are marked by *casas grandes*, the big houses, where the land barons once lived. These can be evocative structures with columns and a wide porch for escaping the scorching Bahian sun. Nearby cows graze in pastures, lined by rows of small shacks, one after the other with a shared wall in between, where the cacao farmworkers used to live. There may be a small garden, and there are always one or more *bagaças*—structures to process and store cacao that slide open to dry the beans when the hot Bahian sun shines and close when the rains come. Behind these farms towers the Atlantic Forest, a bucolic scene that hides the history of this place.

But there is more beneath the surface. The number of bagaças indicates how big the farm once was. Lots of bagaças mean a productive farm, one or two, a smaller operation. While bagaças, like biodiversity records, can be counted, calculated, and translated into literal meaning, other truths here are more elusive. Complex social histories of landscapes are visible only to those in the know, to those who see these places through seasoned eyes.

When we are sitting at a public event hosted by a regional government agency one afternoon, a friend of mine leans over, nods her head in the direction of a distinguished-looking graying man at the front of the room, and whispers, "He was a *coronel de cacao*," a cacao baron. I nod in understanding; we both know this equates to power. Another day, on the streets of Ilhéus, Guiomar lets out a low whistle as she rolls her eyes and points out a young man whizzing past us in a snazzy car. He is oblivious to the people passing on the narrow walkway at the edge of the street and almost mows down a few pedestrians and a dog. She declares, "I know that guy. His father used to be a coronel do cacao, you know. If you look around, you'll see his

Figure 4 Cacao farm with the Atlantic Forest looming behind.

name on buildings, on streets. He has a large family. Everyone knows them." She gossips about how recent years have been tough on them. "His parents died one after another, and they lost many of their farms, but they still have nice apartments here." She motions in the direction of one of Ilhéus's wealthy neighborhoods. Guiomar likes to pass along a mix of stories, gossip, and interpretation. There is truth among them all.

The identification of individuals and families through their historical connections to cacao indicates regional relations, class and race divisions, land distribution, power, and perhaps even future potential. All of this, and more, is wrapped in the seemingly simple statement "their ancestors were cacao barons." This is traced to the fact that harvesting and processing cacao beans at the height of the cacao era demanded a labor force that smaller, family farmers could not absorb. In turn, family farms weren't able to be as productive, and by as early as 1910, most of the family farms in the region had failed.

What rose up in the place of these smaller farms was the *latifúndio* system in which the cacao baron and his family lived in the casa grande and employed hundreds of day laborers to support agricultural production.[17] This division of labor and landownership persisted in Southern Bahia for nearly a hundred years until the late 1980s, when everything changed overnight.

One evening, I sit in a small theater in the city center of Ilhéus. Just steps from me is a bar that the famous Bahian author Jorge Amado frequented in the height of the regional cacao boom.[18] The cobblestoned streets, the majestic buildings that have been carefully restored, the thick wooden floors of the theater, and the carefully painted rooms hearken to an era of riches, grandeur, parties, and wealth for the cacao barons and their families. The actors portray this scene. There are fancy frocks, fountains of liquid chocolate that people stick goblets under to fill their glasses, drink, and refill. A chamber orchestra plays in the background while some dance. Others engage in spirited, unabashed laughing, oblivious to what is starting to happen.

Slowly, from behind a curtain, a witch-like figure appears. She is starkly out of place amid the party, with somber black robes, a contorted face, a stooped posture. Slowly, she takes the large broom she carries and begins to sweep the stage of the musicians, the dancers, and the revelers, all of whom are in her path. The instant her broom touches them, they fall. The chocolate fountain tumbles to the floor with a crash. The witch cackles as she destroys all. The music and laughter subside, and the stage is silent. The local audience needs no explanation. For the older people seated here, in particular,

---

17. Until the late 1980s, the region's agricultural export economy required many low-wage laborers for the *latifúndio*, or large rural estates. These regional farms lie inland in the direction of Itabuna, the largest urban center in Southern Bahia. From the late 1800s to the late 1900s, the economy of Southern Bahia depended on a monoculture of cacao (Alger and Caldas 1994, 108). This system of production created a class of low-wage laborers without the capital to buy land. The cacao monoculture in the region peaked beginning in the 1960s, when the Brazilian government turned to foreign export options and began intensive cacao farming, and "Southern Bahia became the world's second largest cocoa producing area, after Cote d'Ivoire." Cacao was second in exports from Brazil only to coffee, and more than 400,000 hectares were planted in cacao (108). Alger and Caldas also note that large landowners were "both the largest threat and the largest opportunity for forest conservation" (114), which refers to the opportunity to preserve large tracts of forest through Private Natural Heritage Reserves (*Reservas Particular do Patrimônio Natural*, in Portuguese, commonly known as RPPNs).

18. Gilberto Freyre ([1946] 2013) writes of aspects that are well-known in this region of Brazil, namely the *casa grande*, the big house, and the *senzala*, the slave quarters.

the memories of this past are all too present. This scene depicts a mere few decades ago when the insidious witches' broom disease attacked Southern Bahia's cacao crops, wiping out the region's entire economy as well as its social structure.[19]

There are ample rumors of how the witches' broom epidemic came to be, but all converge on one aspect—that the introduction of this disease to the region was far from accidental and that someone, or a group of someones, intentionally brought the witches' broom fungus to Southern Bahia to devastate the cacao industry. Some believe it was leftist political activists who wanted to dismantle the cacao barons, those with land, economic power, and political connections who existed at the top of the region's social hierarchy. Others assert that a handful of employees of CEPLAC (the Executive Commission for Cacao Cultivation Planning), the federal agency that oversees cacao production in the region, were behind this scandal and traveled to the state of Rondônia in the Brazilian Amazon to procure infected cacao plants and carefully transport them back to Bahia by bus in plastic bags stored in the luggage compartment below. "They had a plan," people say, to ruin a few key cacao farmers in the region by distributing cacao with witches' broom onto properties in Bahia with healthy cacao crops.

But things didn't go according to plan; the wind reportedly rapidly spread the disease throughout the region. In a matter of months, nearly all of the farms in Southern Bahia were stricken with diseased cacao crops, and the economic, environmental, social, and cultural landscape of Southern Bahia, which had been completely and utterly defined by cacao for well over a century, was altered forevermore. "The cacao crisis," as it is still called today, transcended class barriers and ruined barons and workers alike. In the early 1990s, more than 200,000 people, much of the region's population,[20] lost their jobs as Bahia transformed from being one of the top cacao producers in the world to a region that *imported* cacao.

Nearly three decades later, Southern Bahia is still marked by the remnants of the crisis—rampant poverty and stark class divisions as well as cacao

---

19. The scientific name for witches' broom disease is *Moniliophthora perniciosa* (formerly *Crinipellis perniciosa*). The popular name for this fungus originates from the fact that the leaves of a diseased plant look like a broomstick, and, as an agronomist and environmentalist friend in Bahia explained, "The 'witch' element connotates the malevolent form of destruction that the disease brings upon farmers."

20. Demeter (1996).

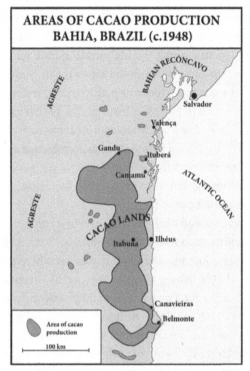

**AREAS OF CACAO PRODUCTION
BAHIA, BRAZIL (c.1948)**

Map 2 Map of cacao region. Map by Jonathan DeVore.

farms that stand idle or have been taken by the government and subdivided for land reform.[21] Memories of this era remain etched in the minds of the elderly and are passed along to the youth. As Southern Bahia struggles to reinvent itself, to find a new identity, I find myself wondering if it can return to become, again, a place "that breathes cacao," as Rui had said. I also wonder what this transformation will truly mean within the individual and collective lives of those living here.

Contesting Space

Almost daily the fuscinha carries me along the main regional road, BA-001, which connects Ilhéus to the fishing and farming communities beyond.

---

21. Land reform will be extensively discussed in chapter 2.

Though the little car is well aged, I push its manual transmission to the limit. Sometimes it performs. More often than not, it rebels against my persistence, and I must stop along the highway and look for a kind soul who can help me out of the mess the fuscinha leaves me in. Sometimes I stuff old rags in its side if the gas cap is loose or pause to pour water over the engine. Other times I am forced to pull over. Invariably, I discover something new, a little store tucked away down a side street, a local bar with chilled beer and a friendly owner. The rhythm of these journeys is emblematic of my fieldwork. Sometimes it is smooth or calls for a small course correction; other times it is halting and frustrating. But once I accept the inevitable, I find beauty and lessons that I did not set out to learn.

Guiomar likes to remind me how, when she first came to Southern Bahia, she walked this entire road that I travel regularly. "I stopped along the way to camp as I needed. Just me and God. It was calm here then. This region was just a series of little fishing villages. No tourists except for the hippies." This makes me chuckle, for a dear friend of mine was one of these hippies. "I hiked from Ilhéus to Itacaré along the beach wearing a skirt and selling little clay figures so I could eat," he recounts. I laugh, imagining his nearly seven-foot frame, made larger by a mop of curly long hair, loping down the beach with a joint in one hand and his wares in the other. Though access is easier now, Southern Bahia continues to attract a mix of tourists, from the same type of hippies just passing through to well-heeled travelers from Brazil's southernmost regions. All who come here must journey along what is now a two-lane highway. While some assert that the road was paved after the cacao crisis to stimulate the economy by increasing tourism to the region's stunning beaches and pristine forests, others claim the real purpose behind the highway was to allow for timber extraction. Perhaps it was both.

On a crisp morning, I arrive in the town of Serra Grande and meet Marcelo, the manager of Bahia's Conduru State Park. The state park was created as an environmental concession with the construction of BA-001 and, to this day, remains the largest state park in Bahia.[22] Marcelo meets me in a white truck with Bahia's state seal on the side marking his official capacity. When he drives this vehicle, he knows that people will know he is coming. Sometimes there are watchmen who keep an eye for him from high on the hill in Serra.

---

22. The Parque Estadual Serra da Conduru (PESC), or State Park of the Conduru Hill, most often called the Conduru State Park, was created in 1997 and is 9,275 hectares.

If they see his truck pass the empanada restaurant far below, a warning will go out to those clandestinely logging inside the park's boundaries, and the loggers know they will have an hour or so to stop and scatter. When Marcelo wants to arrive unannounced, he drives his personal car. While he can't use state funds for gas, he can learn things that help him figure out how to better manage the vast park territory. This is valuable information, for he doesn't live in the park. "A man with a family needs things," he says, "like a school for the kids, a hospital, a place for my wife to work." And so he resides two hours away in the region's largest city, Itabuna, a fact that has, in part, created the situation in which we find ourselves today.

Our truck lumbers slowly on rough dirt roads. It has been raining lately, and the route is worse than usual. Marcelo points out different landmarks that are invisible to me but blatantly obvious to him. A cut on a tree marking a path, likely made by a hunter; bromeliads and orchids growing high above our heads. Suddenly, at a fork in the road, he stops the car, looks over at me, and says in his slow drawl, "You know, I don't really know which are the logging roads and which are the true roads here." He sighs, perplexed. He knows I will likely ask the obvious question about deforestation in a protected area like the Conduru State Park, so he offers an explanation before the words leave my mouth.[23] "I know people take wood out, but what is one guy supposed to do?" he says, referring to the fact that he is the sole park administrator. "I need to get along with the people living here, or they will make my life hell."

He lets the car idle, neither of us wanting to make the call on which road to choose. One leads to our meeting with a community that still lives within the park boundaries, waiting for the state to indemnify them for their properties and help resettle them in a nearby city. The other road leads to an illegal logging area. Both appear to be equally traveled.

Suddenly, Marcelo turns and veers down one of the roads. I don't know if he is feigning recognition or simply taking his chances. Neither of us speak of the possibility of surprising a group of unsuspecting loggers at work. Marcelo spends nights alone in the park each week with no cell reception, no one around, and a sole access road that is often impassable in the rainy season.

---

23. Brazil's National Protected Areas System (Sistema Nacional de Unidades de Conservação da Natureza, SNUC) divides protected areas into different types of conservation areas that range from no use to allowing various forms of sustainable development practices.

And parts of Bahia, like this, can be frontier lands, places where violence is sometimes the swiftest form of conflict resolution, especially when livelihoods are at stake. A few minutes later, we see a group of people gathered in a clearing, patiently waiting for us to arrive and commence the meeting. This day, we have averted danger.

Generating Conservation and Misery

The Conduru State Park is not the only formally protected area in Southern Bahia. Two Environmental Protection Areas, or APAs as they are called, were also created as environmental concessions from the BA-001 infrastructure project.[24] Environmentalists are proud of the region's APAs. They are key to the region's conservation identity and are supposed to allow for both environmental conservation as well as mixed-use development such as family farming and tourism. Rui explains how these APAs came to be. "Our goal was to construct a collaborative atmosphere, a participatory process. *Park* and *APA* are understood throughout the region now," he emphasizes.[25] "Over

24. Under SNUC, Environmental Protection Areas (Áreas de Proteção Ambiental) are state-designated areas intended to unite conservation with limited land use activities. APAs in Bahia have grown tremendously over the past several decades. In 1990, the state had two APAs that covered 13,700 hectares, and in 2020 Bahia had thirty-two. The Itacaré-Serra Grande APA is 62,960 hectares in size with abundant forests, waterfalls, mangroves, rivers, streams, ocean, and coral. The Enchanted Lagoon and Rio Almada APA is 157,745 hectares of Atlantic Forest as well as a lagoon, waterfalls, streams, caves, and mangroves. The first APAs paved the way for other land designations such as the UNESCO World Biosphere status and other important conservation areas such as RPPNs (Reservas Particulares do Patrimônio Natural), or private natural heritage reserves, where scientific research, environmental education, and/or tourism can take place. In 2020 Bahia had thirty-eight RPPNs (see also Drummond et al. 2009).

25. The participatory nature of APA management in this region ideally unites state agencies, local community groups, and civil society actors to establish and manage these protected areas (Oliveira 2005). Many environmentalists in Southern Bahia feel that the participatory management councils that were created around APAs have been successful in garnering widespread support for integrated conservation and development throughout the region. But while Rui speaks of how people have become aware of the *presence* of conservation designations, this has not necessarily created a common perception of the benefits of the APAs or the park. Because of this, a false sense of consensus can overshadow processes of environmental management that derive from the establishment of APAs and parks; though they exist, there is not always a common understanding of why and how local communities benefit. See West (2006)

time, these spaces have been transformed into a public agenda, a common agenda." He pauses, then reflects honestly. "But it's a tough agenda, to create a park, to 'implement an APA,' as we say here. We need to find ways to value livelihoods here, or we're going to be generating conservation along with misery."

The narratives, practices, and perceptions surrounding environmental conservation in Southern Bahia are not singular; they are contingent upon who is speaking and from what perspective. On a hot Bahian afternoon, I stop at a gas station on my way from Ilhéus to Itacaré to have a *guaraná* soda midway in the villa of Serra Grande. I strike up a conversation with one of the region's environmental activists, who also seems to be seeking midday refuge from the heat. He shakes his head when I speak of the road and the resulting environmental concessions, the state park, and APAs, all of which seem to be attempts to marry sustainable development and conservation. "Once the blacktop was complete," he says, referring to the paving of BA-001, "our leverage quickly dissipated for the 'required' conservation actions." He is talking about how environmental concessions go hand in hand with infrastructure projects but often, in practice, remain "unfinished."[26] For example, while the Conduru State Park exists on paper, it also lacks a decent access road and a visitors' center to attract tourists and generate income. But another local man sitting near us overhears our conversation and offers a different perspective on the precarious balance between development and conservation in a global hotspot. He scoffs and says, "Suddenly the APA arrived here, without speaking to anyone. It kind of fell on top of us. People

---

for an extensive study of this tension in Papua New Guinea, Dowie (2011) for a global discussion of a variety of "conservation refugee" cases, and West, Igoe, and Brockington (2006), as well as Brockington and Igoe (2006), for important discussions on the displacement of local peoples for the establishment of parks and conservation spaces.

26. "Paper parks" is a common term to describe conservation designations that exist on paper yet not necessarily in practice (see Banerjee, Macpherson, and Alavalapati [2009]; Reid and Sousa Jr. [2005]). Offen (2003) discusses the making of territories in biodiverse-rich places across Latin America (and Colombia, most specifically), arguing that conservation, land titling, identity, race, and human rights converge, and conflict, around these new territories—a theme that arises in chapter 3 when looking at land titling for quilombo communities. See also Escobar, who writes of how territories involve place-making that is ecological, social, and cultural, processes tied to life itself (2008, 25). When Rui talks about the possibility of generating conservation with misery, he is talking about specific methods of territory-making that have benefits for some and distinctive drawbacks for others.

couldn't do anything that they were used to doing. No one had manioc," he states, referring to a staple farming crop that requires burning of land for production, "and we all had debts." He pauses and adds, "The president of our neighborhood association actually started weeping in a meeting one day because he didn't know what to do. He could no longer farm as he had been because of the APA!"[27] While I have heard environmentalists declare that the APA is too lenient for conservation because it is "a place where you can do anything," others who live here sometimes lament, "The APA is the place where you can't do anything!"

These competing narratives are simultaneously distinct and visible as well as discreet and covert.[28] Driving along the BA-001, signs for the Conduru Park and the APAs are obvious markers of how biodiversity is valued by some. But while territories are denoted by signs, they are also defined—and transgressed—by the discourses and behaviors that uphold or deny the validity of these markers. Those who feel excluded from this public narrative find ways to resist it.[29] When I notice a neatly stacked pile of wooden boards next to a road in the park and ask if there is an actual sawmill in Bahia's largest state park, Marcelo shifts the topic, and his eyes, to evade an answer. He simply says, "People don't always respect the boundaries of the park."[30]

---

27. While APAs encourage certain styles of farming, they discourage others. In the past, manioc has been produced by shifting cultivation (which has been called "slash and burn" agriculture), which isn't permitted in an APA. The *Lei da Mata Atlântica*, or Law of the Atlantic Forest, outlines which farming techniques are permissible in APAs.

28. An example of this evidence, as well as nuance, is DeVore's study of dispossession narratives around a land grab in Bahia by a powerful Brazilian elite, which had important political and economic consequences on a nearby region of Bahia (2018).

29. As Keith Basso (2000) recognizes, any narrative is understood to have several versions. The presence of these apparently different perspectives on environmentalism provides the impetus for questioning the deeper meanings behind people's words, how disjunctures and fissures appear even when an important movement, like environmentalism in a global biodiversity hotspot, is presented as a logical and cohesive whole. I am also inspired by Cruikshank's (1998, 3) emphasis on the power of narrative to uncover rifts, to dismantle boundaries. This is one of my goals throughout this book: to reveal areas of disconnect and fissure with the purpose of ultimately leading to connection.

30. Gupta and Ferguson observe, "Important tensions may arise when places that have been imagined at a distance must become lived spaces" (1992, 11) and, as Brockington (2002) shows, forest conservation is a deeply rooted conservation model that began in colonial times. Scott claims, "The completeness of the cadastral map depends, in a curious way, on its abstract sketchiness, its lack of detail—its thinness. . . . Surely many things about a parcel of land are

One Sunday morning the kids and I travel through Serra Grande, the gateway to the Conduru Park. We stop to get coffee and cake, our favorite Brazilian breakfast. As I struggle to get everyone fed in an effort to keep the arguing among them to a minimum, a man approaches us to chat. It is a slow Sunday morning, and he is clearly bored and perhaps a bit curious about our brood. As our conversation turns to environmental conservation in the region, he winks and says, "You should hang around here on Sundays more often." He nods his head in the direction of the park and lets out a low whistle. I know that on the weekends, Sundays in particular, there is less traffic in the park by the guards who work under Marcelo. "A few weeks ago, in fact, I saw a hunter exiting the park with an impressive catch of twenty pacas!" he declares, referring to large nocturnal rodents. Later, I raise this with a biologist who lives near the park. He, too, tells of a thriving clandestine regional bushmeat industry and says, "I need to be careful where I buy my meat around here." He laughs ruefully.[31]

Respecting or transgressing boundaries, perpetuating or defying particular narratives of environmentalism, are intentional acts that reflect the complex realities of conservation and development in Southern Bahia. And perhaps no entity more than the state is caught in the middle of these dynamics. One day, as I am accompanying Marcelo on his rounds of Conduru Park, we pull up to the residence of a longtime park inhabitant, an elderly man named Seu Auricio. He lives in a *casa de taipa*, a simple structure of mud-packed walls and a dirt floor, which he somehow manages to keep spotless. Despite

---

far more important than its surface area and the location of its boundaries. What kind of soil it has, what crops can be grown on it, how hard it is to work" (1998, 44). The Conduru Park is a "cadastral map" of sorts meant to communicate a specific narrative of environmental conservation through markers, signage, and boundary lines. Yet, at the same time, deeper stories of occupation and boundary transgression remain untold. And as one local man matter-of-factly remarked to me in a meeting on park management, "GIS doesn't mark the people here." For interesting discussions on the role of ethnoscience in conservation to recognize landscapes as lived in with respect to traditional peoples and their livelihood practices in Brazil, see Diegues (2014), and for a discussion of "alternate narratives" that largely favor the presence of a park, see Hartter and Goldman (2011).

31. Silva dos Santos et al. (2020) found that hunting definitely takes place in the Conduru Park. Of the sixty-seven faunal species identified that people hunt, forty-one were captured for consumption. People hunt mainly for consumption but also for medicinal purposes as well as recreation and even "retaliation," supposedly against this conservation designation that limits their activities in this space. See also Castilho et al. (2019) and Teixeira et al. (2020).

a lack of electricity in his home, a sewing machine stands in the corner, carefully covered for protection from dust and rain. Seu Auricio cooks over bricks in a back room, and the bathroom is a mata, the forest. Marcelo takes a seat on one of the few chairs, and I perch at the doorway while they chat about the plans to relocate families, all of whom were living here long before the state park was established. Some families are eager to leave for nearby towns, where they will have better access to healthcare and education. Others know that life in these more urban areas will be vastly different and prefer to continue residing in the park.

The state is in charge of the relocation effort, which requires funding to pay people for indemnification of the lands they will lose.[32] Seu Auricio, who is one of those who wants to leave, is frustrated it is taking so long. At the same time, however, he and his neighbors are technically prohibited from planting crops for the coming year. They are living in a state park, after all. As Marcelo delivers the news that the funding and indemnification process still hasn't come through, Seu Auricio looks at us, sighs, and says, "I guess we need to burn the ground to prep it and plant again. There is no other option." Marcelo nods almost imperceptibly but doesn't counter his statement. While Marcelo has declared publicly he will defend conservation in the park "até o exército," meaning to the point of bringing in the army, he has also admitted privately, "How do I tell a hungry person living in the park that he can't feed his family?" And when the state is absent or unable to implement conservation, other entities try to step in to help.

Existing with the Kings of the Black Coconut

The NGO Yonic has taken over the small town of Itacaré. With prime real estate on one of the town's main beaches and a rustic-chic office made of recycled bottles and low wooden couches, the new organization is trying to do good things in a place that sorely needs it. For a while, Yonic assumes

---

32. Within this park, there are a number of private property owners as well as *posseiros*, squatters, who have resided here for years. The entity in charge of park management, the State Secretary of Environment and Hydrological Resources, has begun indemnifying the posseiros' land and relocating them for a modest payment. However, given the cost of buying out the large farm owners who reside inside park boundaries, as of 2009 (and still to this day) about half of the park is still in private landholding of mostly middle-sized farms without workers.

what would normally be municipal responsibilities such as collecting the city's trash and recycling and taking it to a landfill out of town to sort and repurpose it if possible. Yonic holds environmental education classes and works with youth in some of the town's poorer neighborhoods in an effort to mobilize them around the environmental cause. It even sponsors surfing competitions, distributing marketing propaganda far and wide that claims, "The event site is as much a paradise as the region itself; our award stage is made of recycled plastic bricks, coconut sheath paneling, abandoned wood fencing, chairs and benches from woven recycled newspaper, and natural waterfall showers created from small forest streams."

This is the public face of the regional NGO, and for a while the excitement around Yonic is palpable. Privately, however, some of the NGO staff themselves express doubts about their work. They worry about their ability to make change here. This is a foreign NGO, and there are a lot of complex cultural and bureaucratic dynamics to contend with. As a common saying goes, "Brazil is not for beginners."[33] Bahia may be the epitome of this sentiment. As we sip beers on the front porch of Yonic's office, located on one of the town's main beaches, one of the directors complains, "The people here just don't have the capacity to understand." This statement hangs between us, as if he assumes it is clear to me, but I myself feel like a beginner in this murky landscape of meaning and am unsure how to counter what seems like a disparaging remark.

A few days later, I visit Yonic's founder back in the beachside office. He is an old man with a colorful reputation that precedes him. Some say that he was a revolutionary during Brazil's military dictatorship era and that he has come here to hide out. Others believe he is the leader of an eclectic group of young idealists who view him as a guru of sorts, calling him "Father" and honoring his every wish. Yet others assert the NGO was actually founded and supported by a young woman from a wealthy Swiss family. No one really knows. All of this, or none of it, may be true. As our chat comes to a close, the NGO leader looks at me and offers his assessment, which aligns with that of his staff: "People around here, you know, still really like mirrors." Though he offers no more explanation, I wonder if he considers the work he is trying

---

33. This quote comes from the famous Brazilian musician Antônio Carlos ("Tom") Brasileiro de Almeida Jobim (January 25, 1927–December 8, 1994).

to do as liberatory from the colonial forces who came and tried to "pacify" locals with gifts of shiny objects.[34]

This leader is not the only one who shares the perception of the difficulties of coming to a place that is still trying, sometimes precariously, to break free of its colonial legacy. The next day Rui and I finally find a chance to escape the chaos of his office, which is always packed with busy staff bustling about, phones ringing, and people coming and going. We sneak out to have a quiet lunch together. At the end of our time, I pose a pointed question: "Tell me, what are the greatest challenges you confront in your work?" Without hesitation Rui responds with a statement that surprises me: "One of the greatest challenges, my friend, is that sometimes people here can still perpetuate attitudes of colonialism."[35] As we walk outside the restaurant, a tall Afro-Brazilian man, who appears to be about eighty, sells sugarcane juice that he makes with a rough hand grinder. A few raggedly dressed barefoot boys offer to help people park and then to "watch" their cars while they shop or dine or conduct business. Beyond this scene lies the Atlantic Ocean, where the Portuguese and the people they enslaved first arrived more than five centuries ago. Aspects of the colonial legacies of race and class remain starkly imprinted in the region today.

After a long day with Rui in Ilhéus, I drive home, ready to see the kids and begin our nightly routine of a neighborhood walk, dinner, showers, and bed. They are full of energy, though I am spent. "I did a *macaco* in capoeira today!" Ella says excitedly, referring to a complex move where you balance on one hand upside down and cross your legs in the air. She is the capoeira master of the family. Not to be undone by Ella's capoeira prowess, Max exclaims, "We

---

34. In this comment, he was referring to the history of how Portuguese colonizers reportedly tried to conquer Indigenous peoples not only by force but also by offering shiny objects like pans and mirrors that were new and exotic to "pacify" the people living in the region they invaded.

35. Environmentalists in Southern Bahia are sometimes those who are the *most* aware of how their work can be beneficial and instrumental yet, at the same time, can also perpetuate historical power inequities. There are myriad relations surrounding conservation, from ideology to the need to adhere to the requirements of donors and funding organizations from outside the region, which sometimes can mirror colonial practices of the past (see Sundberg 2006). At the same time, NGOs also play a critical role in Brazilian conservation, especially considering the limited capacity and funding of the state (see Fonseca 2003; Hastings 2011; Mittermeier et al. 2005). See Koslinski and Reis (2009) and Pearce (2010) on how NGOs are important sites of local-to-international politics.

got pet turtles—come see! We named them Barbie and Tom." Indeed, two turtles slowly crawl about and munch on lettuce in a small plastic tub in the kitchen. Guiomar looks at me proudly. I wonder if her boyfriend, Manoel, found them in a nearby pond. There are no pet stores here. To calm the kids we set out on a family stroll. As we enter the main street of town, we run into a friend along our path, a spunky *nativa*. She is headed to a capoeira class near our home and falls into step with us to walk together. After hearing the kids' news from the day, she politely asks, "Hey, how's that work of yours going?'" I tell her about the latest meetings I have been attending, the people I have chatted with, how I spoke with the head of one of the prominent regional NGOs today. She listens and then retorts, "When people have an NGO here, they think they're the *rei da cocada preta*," the king of the black coconut, which simply means "a big deal."[36]

NGOs have proliferated over the past few decades, not only in Bahia but throughout Brazil.[37] "We have a lot of NGOs for such a small region," people often remark of Southern Bahia, a fact that initially attracted me here. When the state is absent or fragile, as it often is here, NGOs can step in to act in ways that are both instrumental and paternalistic. While Rui often reminds me that people here have started to view NGOs "as the state," meaning they are often left to handle conservation largely on their own, my local community leader

---

36. She is referring to a popular samba song that talks about the most powerful or attractive man as being the "king of the black coconut sweet," meaning he gets to feast on the tastiest, darkest *cocada* (a sweet made from coconut and sugar), which is both a racialized and gendered comment too. In the commentaries and judgments that often rise up in both reaction and relation to the discourse and practices of biodiversity, below the surface lie sites of struggle and negotiation. I have heard environmentalists branded *novos coronéis*, or "new colonels," who work for *rei ONGS*, or "king NGOs," revealing perceptions of environmentalism not as a liberation movement that will critically affect the future of the natural and social environments in Southern Bahia but instead as a form of domination that emulates the colonial past. See Braun's (2002) ethnography on how tensions among colonial power, forest politics, and struggles over meaning are exerted onto forested landscapes like Canada's Northwest Coast to produce both present-day histories of colonialism as well as efforts to produce a postcolonial future.

37. The environmentalism that has developed around the Atlantic Forest is most often expressed and perpetuated through the environmental NGOs that are rapidly proliferating throughout this region. Over the past three decades, NGOs in Southern Bahia have grown stronger, more organized, and more numerous. In 1990, there were a handful of environmental NGOs; in 2009, more than fifty; and in 2021 closer to seventy-five. This growth mirrors the burgeoning of NGOs around the world, "a revolution as significant to the late twentieth century as the rise of the nation-state was to the nineteenth century" (Salamon 1994; see also Petras 1997).

friend Catu has another take: "A majority of the NGOs in this region are *picaretas*," meaning fake or false. Yet another leader in the region, a young man named Bruno who moved from São Paulo and became the right-hand man of one of the city's most financially well-endowed and active NGOs, explains, "In Brazil there are five million NGOs—but only one million are worth their weight." People often comment that you can buy the support of an NGO to get an environmental stamp of approval on whatever project you happen to need it for—from skateboarding competitions to musical events.[38]

At the same time, however, NGOs can represent a new type of activism predicated upon citizen involvement and community empowerment. While federal and state agencies often lack the necessary funding for personnel, vehicles, and even gas to travel around Southern Bahia, NGOs can be more agile. Ideally, they are driven by a strong mission in a region that, with the fall of cacao, desperately calls for one. "I'm going to start an NGO" is frequently heard around here, showcasing the dream, the possibility, and the agency that can align with ambition. And if you don't start your own NGO, working for one is the next best thing. Upon one of my regular departures from Brazil, Guiomar poses a pointed question. "Isn't there any way you could help me get a job with an NGO? I want to make something of myself, to do good work

---

38. The ways in which environmentalism is constructed and perpetuated are neither singular nor static. In one sense, environmentalism is expressed through biodiversity statistics and practices that flow through the designations of parks and reserves, the funding of environmental projects, and the like (see Igoe [2017] for a rich study of how "modes of representation" from charts to diagrams to statistics can mark conservation spaces, funding streams, governmental progress toward commitments, and levels and types of engagement with communities). But the environmental movement is also shaped by *reactions to* biological discourse and conservation practices. Underneath their breath, people sometimes talk of environmentalists and conservation NGOs as "controlling the region" or of how friends or neighbors are "bought by NGOs" if they partake in environmental projects. One community leader calls the people throughout Southern Bahia "victims" of NGOs, while another laments, "It is very difficult for people to speak out against NGOs." At times in Southern Bahia, these cultural politics of difference are obviously identifiable, and at other times they are nearly imperceptible yet still present in more subtly hegemonic ways. In some cases, people may show up at a meeting but feel uncomfortable speaking, even when they have an opinion on a conservation proposal. In other cases, resistance is shown more overtly through tactics that deny the validity of or openly defy the biodiversity narrative.

while you are gone." Nodding vehemently, she continues to try to convince me to recommend her. "I believe in this 'environment mission,' you know!"[39]

Differing perspectives on environmentalism converge, conflict, and even coexist. Here the value of biodiversity, as well as legacies of colonialism, continually vie for power. As always, Rui has a philosophical take on these complex regional dynamics when I raise them to him. "There is a *papo novo*, a new discourse, beginning here," he reflects one day as we walk on a forested trail along the coast. "And to be honest, we participated in the imposition of this papo novo—the imposition, more than the construction—of an environmental agenda in the region, an agenda built on biodiversity, *on* the Mata Atlântica."[40] What he is talking about is the delicate balance between imposing conservation and establishing the necessary processes and practices to have it rise up from the people who live in the region. "We thought, we need discipline here. Let's comply with the legislation. And as this region had a culture of informality, of transgressing the principles of the constitution in many areas, and still does today, attempts to construct an environmental agenda independent of legislation, they were always weak here." He concludes, referring to the need for formally designated conservation areas, like parks and APAs, places that adhere to Brazil's Forest Code. For this is a

---

39. Brazilian socio-environmentalism was born in the late 1980s as a result of interactions between movements that were overtly social in nature, such as the rubber tappers movement in Amazônia and the land reform movement throughout the country (Keck 1995; Schmink 1982), alongside, but different from, the more biologically focused global environmental movement (Santilli 2005, 31). As environmentalists joined traditional populations such as Indigenous and river peoples (Santilli 2005, 32), "environmentalism allied with the social movements, and social movements turned more environmentalist" (Crespo 1993, cited in Santilli 2005, 51). To this day, social movement leaders often classify themselves and are perceived by others in hybrid and overlapping ways. Because of this overtly social-environmental connection, throughout this book I conceive of environmentalism as a broad social movement. Furthermore, farmers and land reform activists also consider themselves environmentalists, as do many nativos, while people who are environmentalists or farmers can also identify as nativos if they were born in the region. These categories are not singular; they are shifting and often hybrid. See Conklin and Graham (1995) for a discussion of this in the context of international environmentalism, social movements, and the eco-politics of Amazonian Indians.

40. Studies of international NGOs working in Brazil have shown how these organizations can be instrumental for translating science and stakeholder participation into important policy changes (see Hastings 2011).

Figure 5 Rui (right) talks with a local farmer about integrating conservation and development in the region.

region with a paucity of remaining primary forest and it *still* has incredible biodiversity.[41]

Rui raises the million-dollar question for Southern Bahia: "The challenge is to try to *construct* an environmental agenda in the middle of an economy, a society, in crisis." Rui and I both know that the Biosphere Reserve, the state park, and the APAs provide the foundation for new economic options, for new visions of livelihoods that are based on the presence, rather than the absence, of the forest. But they also demand buy-in. Everyone here must

---

41. Considering biodiversity conservation and greenhouse gas emissions reductions—globally important goals—there is no denying the importance of formally designated conservation areas of the Atlantic Forest. Also, as Ribeiro et al. (2011) note, "today's total protected area of the Atlantic Forest is approximately 2.26 million ha, which represents only 1.05% of forests of the original cover distribution, way below the 10% recommended by the Global Strategy for Conservation. . . . Nature reserves protect 9.3% of the Atlantic Forest remnants, however, differently according to the regions" (2011, 424). Only 4.2 percent of the Bahian Atlantic Forest is under protected status, and enforcement of the Atlantic Forest Law is lax due to limited state capacity (Ribeiro et al. 2011, 424).

run after, or at least run in the same direction toward, a common vision and agenda for the region if these places are to remain intact and to avoid, as Rui had said earlier, generating "conservation alongside misery."

Rui and I continue our hike and reach the top of a hill that faces east. The ocean far below is not far from the site where Brazil was first colonized; the "Discovery Coast" lies along the coast to the south. We both stare out at the ocean, feeling a bit lost in the complexity of our conversations.[42] Then, way out in the distance, we notice them at the same time: a pod of whales surfaces, migrating south, swimming in a powerful, gentle dance of unity. We depend upon moments like this to remember the wonder of where we are and the importance of finding ways of coming together, of altering the outdated narratives and practices around environmentalism and creating new ones that apply to the contemporary regional context.[43]

Partnering on Production

I toot the horn, emerge from the fuscinha, and gingerly step along the narrow path leading to the home of Beco and Eliana, trying mightily to sidestep puddles of mud in my flip-flops. I take a seat on the plastic-covered couch in the living room of their modest but inviting three-room home. As always, I am amazed by its tidiness. The first thing anyone walking into their house notices is an impressive television and stereo on full display in their living room. When I point out that there is no electricity running to the house, Beco has a quick

---

42. Brazil's Forest Code and National System of Conservation Units are essential for environmental conservation. The Forest Code in Brazil requires that all landowners with forested land must set aside a percentage of this land as a legal reserve. While these legal requirements are an essential foundation for conservation, sometimes family farmers and other populations living near the forest understandably conceive of environmental regulations and laws as constraining their livelihood options, which can create an ideological lightning rod that can separate people in profound ways (see Drummond, Franco, and Ninis 2009).

43. Runk (2017) writes of alternative narratives that shape environmental governance. In a related vein, I am curious about *alter*-narratives that differ from (but do not necessarily outright counter) dominant narratives. While counternarratives imply that there are obvious and opposing narratives that arise in resistance to dominant narratives, alter-narratives propose that there are often less obvious, yet still important, changes taking place in relation to dominant narratives. Alter-narratives can be subtly optimistic efforts to dismantle the hegemony of dominant narratives, as we will see in the cases on the concluding pages of this chapter.

answer: "When it arrives, I'll be ready!" Carefully placed family photos hang on the wall, and a pot of hot coffee either awaits or can be put on at a moment's notice. These are people who are prepared for whatever the day brings.

It wasn't always like this. Eliana talks fast as she tells their story. Her voice shakes with emotion, as she remembers this period well. "After months of a miserable living situation," she explains, "I took matters into my hands. I left ahead of him." She glances sidelong at Beco, and then at me as if we both know that sometimes women need to take the lead. "I had to do *something*." She rolls her eyes and continues, "So I wrote the date that I expected him to join me on the stone wall in front of the tent we were living in, set off on my own, and came here." She motions around the room. "But it wasn't like this back then. I was sleeping on the bare floor, crying myself to sleep every night, and six months pregnant with a growing belly!"

Figure 6 Beco and Eliana with one of their goats.

I take in her slender frame, not more than a hundred pounds after having her two kids. She continues, "Yeah, that was a hard time, but people helped me. One day my mother loaded a new cushion on the bus from Ilhéus and then carried the mattress on her back all the way from the bus stop to here!" Eliana looks at the long, rough dirt road used to access their property.

As the story goes, Beco soon came to live with Eliana for good. They planted crops. They had the baby. Gradually, things began to improve. But they were far from comfortable. Eliana recounts, "Fortunately, I produced a good amount of milk to feed our child as we didn't have money to buy products from the market." Beco interjects, "But we were fighters." Eliana nods vehemently. They had another baby and continued on, running after their efforts to farm, to make their lives here. And then one day, staff from the NGO that Rui started, Floresta Viva, came by to ask if they were interested in joining a project growing organic herbs, lettuce, and arugula, as well as raising goats and chickens. The NGO folks guaranteed that nearby ecoresorts and pousadas that dotted the region for tourism would provide a market if they adhered to the rules of organic farming and stopped using pesticides to farm. They even promised regular technical assistance. "We had little to lose," Beco explains.

In just a few years, Beco and Eliana became one of the most successful stories in the region. Now their house is a regular showcase for visitors, an example of successful small-scale, organic agriculture on a case-by-case, family-by-family basis.[44] They work hard by selling their produce to the nearby ecoresort through a weekly organic fair for farmers in the region.[45]

---

44. Piasentin and Saito (2012) argue that agricultural diversification is essential for managing the risk in agricultural production and that other crops should be produced alongside traditional cacao production using agroforestry systems (*sistemas agroflorestais*, commonly called SAFs) for maximum biodiversity conservation and sustainable development.

45. Salaries in Brazil are measured by multiples of a monthly minimum salary. In January 2009, when I spoke with Beco and Eliana, the minimum salary was R$465 (then US$200.95) per month, and at the time of this writing in November 2021, it is R$1,100 (US$201). In this case, environmentalism crossed from a narrative of strict conservation to encompass opportunities for family farming on already-deforested lands in an effort to keep the Atlantic Forest intact. In a shift from the previous domination of a large-landholding cacao colonel, narratives of collaboration are now arising. Beco and Eliana, as well as other family farmers, are "partners" with Floresta Viva, "members" in the organic agriculture association that runs the weekly fairs, and "companions" of the nearby ecoresort that regularly buys their produce. These arrangements run counter to environmental conservation as contemporary colonial imposition; for these

When I return, Beco and Eliana call to invite me over to visit. They can get a cell phone signal if they walk up the road a bit. Though electricity still hasn't arrived at their property, their neighbors across the street have it. It's close, says Beco. For him and Eliana, and many others, hope springs eternal.

Altering Control of Cacao

The cobblestoned streets of Itacaré are packed with tourists in town for a holiday. People amble slowly to window-shop, a few folks scurry to participate in a capoeira circle, couples pass by on their way to an evangelical church service with Bibles tightly clutched at their sides, while others sit out on the steps of their business giving fist pumps and calling out to friends who pass by. A well-appointed store in the middle of the street sells chocolate truffles filled with every imaginable delight, from coconut to *cupuaçu* fruit. On the wall advertisements lure tourists to tour a farm up the river about an hour away where the cacao is grown for the chocolate. Amid the cacophony of the nighttime bustle that characterizes Itacaré, a lone voice sings out, steadily approaching, getting louder by the moment: "Chocolate, chocolate of Bahia, come to taste the chocolate of Bahia, you won't regret it." A slight young man limps along the road on severely deformed legs, pushing a cart, and singing at the top of his voice, which is quite good. He stops and gives us a hand slap greeting. We purchase some chocolate truffles, and he moves on to the next potential customer, his melody echoing off the walls of the homes and businesses that line the narrow main street.

When we first traveled to Southern Bahia in 2003, there was no good chocolate to be found in Brazil's leading cacao-producing region. To be sure, milk chocolate was abundant. Though the neatly arranged case of chocolate sweets lured unsuspecting visitors through marketing as "authentic" chocolate from Bahia, the added ingredients of sugar and milk masked the taste of

---

arrangements to work, environmentalists and family farmers need each other. Environmentalists need farmers to adhere to conservation regulations in APAs and parks, and farmers can benefit from access to capital, markets, and technical know-how in the best practices for organic farming. Reflecting the rollback of the state, while CEPLAC used to provide technical assistance to family farmers in the cacao era, when the cacao crisis happened, NGOs and private businesses like the nearby ecoresort became the only entities occupying this role, and at the same time, these businesses depend on farmers to feed the growing demand for organic produce by ecotourists in the region.

the bean as well as its health properties. Furthermore, there was no regional *fábrica* or chocolate-producing factory. People often lamented that the best cacao beans were exported to the South of Brazil.

As Rui had explained to the hopeful guides-to-be in the training workshop, Southern Bahia has forest today in large part due to its long history of cacao; growing cacao depends on the forest overstory to provide shade for the cacao trees on the floor below. Because of this, *cacao cabruca*, often called simply cabruca, an agroforestry technique of planting cacao under the shade of native vegetation in the tall primary and secondary Atlantic Forest, has existed in Southern Bahia for over a century.[46] "What's going to save this region are the cabrucas," says a prominent biologist who has made her career here. The cabruca method of farming is not only historical; it is also cultural. Researchers have even found that when farmers are incentivized to *not* use the cacao-cabruca shade-growing process and could possibly produce larger yields by using chemicals, they still often stick with what they know, which is cacao cabruca. But at the same time, biologists also alert that cabrucas are declining due to regional land use changes. But how can this region safeguard its cabrucas? And what benefit will this bring, and to

---

46. Traditionally, cacao has been grown in monoculture (though under the shade of various other species). But various crop combinations involving cacao have recently been undertaken by the farmers with encouragement from the Brazilian government (Alvim and Nair 1986, 3). While cacao is associated with eras of boom and bust based on worldwide production and markets, twenty years ago the nature of production changed drastically because of a combination of a market crash and the witches' broom disease. Today, cacao is most often part of a larger agroforestry system, and Southern Bahia is home to cacao cabruca—an integrated system where the cacao is cultivated under the native trees of the Atlantic Forest. This system was introduced to the region in the late eighteenth century, and today it is disseminated across more than fifty municipalities. According to the Worldwatch Institute, "in southern Bahia, only around 4,200 square kilometers of original forest remain—cabruca has become, by default, a major conservation asset. The region's cocoa belt covers some 13,400 square kilometers, of which 4,800–6,000 square kilometers is actually planted in cocoa, and most of that planting is likely still cabruca. Although many other countries grow some cocoa under native shade, Brazil has the largest 'cocoa forest' in the world" (Bright and Sarin 2003, 27). Ecologists advocate: "Biodiversity conservation strategies should include the promotion of productive yet biodiversity friendly cocoa farming practices (land sharing), associated to the expansion of protected areas and assistance to protect legally mandated on-farm reserves and create voluntary private reserves (land sparing)" (Nogueira et al. 2019, 1962). At the same time, while some local actors advocate for expanding organic management in cabruca systems (Carvalho and Barbieri 2013), organic cabruca farming has a high labor cost, which can be a barrier to entry for farmers wanting to make this transition.

whom?[47] I need to understand the flow of cacao. To trace the cacao supply chain in this region from farmer to export, I seek answers in Ilhéus, where raw cacao beans converge from the regional farms where they are produced and are then exported beyond Bahia.

I weave through the streets of Ilhéus and take a shortcut to avoid the increasingly congested traffic in the city center. The road winds away from the beach, away from the town's port, and passes big industrial-looking structures marked by the names Cargill, Bunge, Barry Callebaut, the largest buyers of cacao in the region. These companies purchase raw cacao beans for export to the South of Brazil, or internationally, so that the beans can be turned into chocolate products. In most cases, no one knows the source of the bean nor what sourcing from the South of Bahia really means, anyway, in terms of a product differential. I'm on my way to an appropriately named cooperative, Cabruca, which often works closely with these larger companies.

I arrive at Cabruca's office, located at the top of the highest hill in Ilhéus. I enter, meet the head of the cooperative, and sit down in his well-appointed room for a lesson in the cacao supply chain. "Most of the cacao being exported isn't directly from the individual family farmers. It goes through our association, Cabruca, or sells to Cargill," he begins. For three hours, we talk cacao production, export, pricing, value-add. There are many steps between the farmers who produce cacao and the finished product. But I think back to conversations with various family farmers in the region who

---

47. Although biologists recognize that cabruca does not produce the level of extreme biodiversity found in an untouched primary forest ecosystem, they agree that cacao cabruca is still the most sustainable form of farming and should be promoted for the future of the region. There is extensive work on cabruca (see Araujo et al. 2019; Cassano et al. 2009; Faria et al. 2006, 2007; Nogueira et al. 2017; Rolim and Chiarello 2004; Sambuichi and Haridasan 2007; Schroth et al. 2011).

It is also interesting to consider, though they could sometimes make more money with chemicals and clearing in the short term, farmers can be a conservative lot. In Southern Bahia, for example, they assert that the shade of the forest helps the cacao plants grow (see Johns 1999, which shows how farmers reject higher-yield options for cacao production because they perceive these systems as riskier). Farmers often employ a diversified agroforestry system, or *sistema agroforestal*, also known as an SAF; they grow cacao alongside other crops such as bananas, palm oil, coffee, yerba mate, nuts, açai, and umbu. Some also diversify their income by harvesting rubber, raising small animals, or collecting bee pollen. Like Beco and Eliana, they don't bet on a single income-generation strategy. Rui asserts that 80 percent of farmers with whom Floresta Viva works want to implement SAFs and that 100 percent of these producers view SAFs as important for integrating farming and conservation.

must go through *atravessadores*, or pass-throughs, that could be multinational companies like Cargill or regional associations like Cabruca. They are the reality here, and changing a reality is hard to do.[48]

One afternoon I take shade under the Atlantic Forest, walking slowly through a cacao farm with my farmer friend Gilberto. He is a jolly man of about fifty who has been farming most of his life. He used to work for others, for large landowners, but now, with the cacao crisis, Gilberto has managed to get a small plot of land to call his own.[49] I wonder aloud why he needs to send his product through a middleman, why he can't have more power as the one who grows the regional product that is starting to make a comeback with each year that the witches' broom disease becomes an increasingly distant memory. "Well, it isn't as simple as you think. We don't have direct connections to exporters," he elaborates, patient with my apparent ignorance and naïveté. "We need this access," he emphasizes. I nod and remain silent, wondering to myself whether farmers recognize the opportunity the agricultural conglomerate or a cooperative can afford them and are willing to overlook, or de-emphasize, the deeper questions of fair compensation, access, and control surrounding partnerships with these atravessadores. They arguably have more negotiating power if they band together in associations; organization and unity can be key for overcoming power differentials.[50] I also wonder

---

48. While large companies control a variety of the region's cacao export, they can also serve as important connections for markets that farmers alone don't have. Some farmers lament having to go through these companies, while others see them as adding value. As with many things in life, it all depends on one's position.

49. When the witches' broom disease affected the region's primary crop of cacao in the late 1980s, large landowners often had to abandon their properties or convert them more intensively to cattle ranching or coffee (Saatchi et al. 2001), while small properties were able to mitigate the impact of the disease. According to Brazil's Forest Code, all privately owned properties are legally bound to keep 20 percent of their land in protected reserve status, where logging and agriculture are not allowed (Chomitz et al. 2005). Larger tracts beyond this legally required 20 percent (which can be Private Reserves for National Patrimony, RPPNs) are also important for conservation. Rezende et al. (2018) outline how compliance with the Forest Code, which includes restoration of degraded areas by 2038, could increase the Atlantic Forest by up to 35 percent, turning it from a hotspot to a "hopespot," as they call it.

50. Colonial relations of labor have been transformed to contemporary forms of work for people living in a biodiversity hotspot, resulting in particular notions of forest value (see Sodikoff 2008, 2009). This environmentally related alter-narrative of sorts reveals a shift from the past production of cacao: it is no longer a monocultural crop and is regulated to be environmentally friendly through a system that stresses biodiversity through an organic cacao-cabruca growing process.

if contemporary colonialism still converges around cacao—around access to markets, control over prices, being the one in the middle to do the deal—or on the other hand, if perhaps farmers here are pragmatically recognizing the limits, as well as the possibilities, of participating in global markets. After all, one of the biggest advantages of producing organic cacao is that it fetches a higher price for the producer—and who wouldn't want that?

Just as I am thinking this through, though, Gilberto picks up a cacao pod and throws it to me to bring home to the kids. He also throws me for a loop with a statement that reveals he is a few steps ahead of me in analyzing the complex historical and contemporary relations of cacao production in Southern Bahia. "You know, what we really need is a processing plant right here in the region, right here and now," he states. "*That* would give us the access, the power, we really need." He winks. I also know that to make this happen, the region depends upon strong local-level leadership, upon those who are able to perceive the many ways that history, culture, and positionality shape people's conceptions of and actions toward the region's natural environment.

Continuing a Leadership Legacy

I drive with Rui to a meeting he is having with farmers in the region. He has been invited to give a presentation on how a legal reserve is required of all property owners possessing land with remaining Atlantic Forest. He drives cautiously along the dangerous two-lane highway, navigating the bikes, pedestrians, and stray dogs veering in and out of the narrow roadside in a Zen-like way. He is quite different from me—he is careful, thoughtful, deliberate. As we make our way, I ask him about *his* story, about how he has come to be involved in the environmental movement and identify himself as an environmentalist or, more accurately, as a social-environmentalist.

"In the 1970s," he begins, "Brazil started to discuss the environment. Though I was still a boy, my involvement in this cause started directly with a relation to nature. Then my older brother became an environmentalist, working with NGOs while I studied agronomy in the 1980s." He pauses and further explains this connection, "Agronomy in communion with nature, from the perspective of conservation, not from the perspective of intensive use," and then continues explaining his life's trajectory. "I was also involved in the student movement. I went to the Amazon, I worked with IMAZON, an NGO in Belém.[51] This was a strong experience in my life." As we drive past a small farm, he looks over at it and offers, "Then I started to study and ended up learning about leaders who were rural farmers, their discourse, their vision. It was really cool." He pivots to his national and even international connections, connections that, to this day, fuel the NGO he started. "After doing my master's in Rio in public policy, I came back to Southern Bahia to work in this area. When I was at IMAZON, people from Conservation International visited me. They were thinking of founding an NGO in Southern Bahia... me, already being a Bahian, already knowing Amazônia, already knowing financing agencies that were active in Amazon and were also active here...." He trails off, leaving the obvious unstated—that training and knowledge and relationships set him on the course to where he is today. For a moment, we drive in silence. He concludes, waving his hand from the forest on one side of the road to the ocean on the other, "And this is everything that I wanted. I wanted to return to my land and work with *this.*" He looks at me and nods with certainty.

Suddenly, I draw a connection among details, facts, and experiences that I have forgotten, memories that have grown fuzzy and remained dormant in the back of my mind. Until now. Rui is the ultimate listener, philosopher, and seeker, always trying to connect with people, with what matters to them. On that night back in São Paulo, after Mario made a phone call to introduce me to Rui and arrange for us to meet, I had started to head out into the chilly night. "Hey!" Mario yelled as I reached the door, calling me back in for a moment. I turned and looked at him, wondering what he had forgotten to tell me. Mario looked at me intently. He wanted me to capture what he was going to say. In a voice that was uncharacteristically solemn, he said, "Rui is *gente boa*, good people, you can trust him." I remember looking at Mario, nodding, still not understanding the deeper meaning behind his words. He paused, then continued. "Remember that night in Salvador? That night we went to the boat of our friend Pedro, the environmentalist who died while diving too deep?" I nodded again. Then he said what I needed to know and what I connected now: "Rui is Pedro's brother."

---

51. Brazil's environmental community is surprisingly tight and cohesive, considering the size of the country; everyone seems to know each other. IMAZON (Institute of Man and the Environment of Amazônia) is a well-known Amazonian NGO headquartered in Belem.

Narrating Environmentalism in a Biodiversity Hotspot

How do narratives reveal how social movements are constructed, perpetuated, and validated, as well as resisted and reconfigured?

While environmentalists perceive this region through a lens of biodiversity, constructed upon scientific knowledge, facts, and statistics, those who exist apart from or alongside Southern Bahia's environmental movement can view the environmental movement as a form of domination and control. This chapter explores how competing narratives of environmentalism are constructed, perpetuated, resisted, and recast. It examines how the colonial past and the global emphasis on conservation in the present shape Bahia today. It challenges us to perceive and analyze how different perspectives on environmentalism converge, conflict, and sometimes coexist in often unspoken yet powerful ways, with the goal of better understanding the dynamic environmental movement in Southern Bahia, a biodiversity hotspot.

Striving for Land and Livelihood

"I work on the farm, on the land . . . working with my hands, trying to produce the dream if it is possible."

—Maria do Carmo, Itabuna, Bahia

Traveling Inland to the Farms

Some have aptly described the Atlantic Forest as "a landscape scarred by human striving."[1] The nearly two-hour drive between Itacaré and the State University of Santa Cruz, simply known as UESC, is a visceral reminder of this. Signs for state parks and protected areas indicate how environmentalists throughout the region are working to implement conservation practices. People walk barefoot along the region's main highway with bundles of downed wood on their heads, revealing the manual labor required to feed their families who live in homes for which they cannot afford gas. Remnants of coastal resorts abandoned mid-project to decay in the hot Bahian sun hint at histories of developers who once aimed to build a dream predicated on Southern Bahia's natural beauty. And some of the most pronounced accounts of human striving in the Atlantic Forest are seen in the lives of family farmers, the people who live inland from the coast trying to carve out a place for themselves every day by making a living off of the land.

Leaving our small beach town is the easy part. I always start the journey with energy and ambition. Once in a great while I strap a surfboard to the

---

1. Dean (1997, 1).

top of the dilapidated fuscinha in the hope that on the way back from my fieldwork I will find an hour to "fall in" the waves, which I manage a handful of times. Most often, however, I drive past the beaches, speeding home to the children, to a different type of work, through attempts to come back from someplace else, to be an engaged mother for a few hours, and then, late at night, to return mentally to the place I was earlier physically. This is a strange, exhilarating, frustrating existence. I am often wishing I can find better ways to fall in, in all ways, to where I actually am.

Travel to UESC requires going south along the coast toward Ilhéus. At the edge of the city, men stand on the narrow strip of sidewalk selling massive fish they caught earlier in the day. One man often dangles the fish as cars speed by while his friend stands ready with a Styrofoam cooler for any prospective buyer to take the fresh catch home. At the north end of Ilhéus, I turn westward, inland, and drive past the multinational companies strategically located near the city's port to export cacao beans around the world. For centuries, this has been the development model.

Driving past the federal police station on the way out of Ilhéus, I always turn my head to look at the vacant courtyard outside the entryway. I remember a day long ago, symbolic of much of our time in this region. We had come to the police station to carry out a bureaucratic procedure that had to be done regularly—registering our presence within thirty days of arrival. As always, things had not gone smoothly because the police had determined that we had done something wrong when in fact we had not. After a long, hot drive to the station and hours of waiting for our appointment, only to receive the unwelcome news that we were somehow violating a supposed law, tensions were high. "This is insane," Jeff had yelled loudly in English to a group of people who stared blankly at him, stamping his foot and rolling his eyes. To him, there was nothing worse than the suspicion of corruption and the inability to arrest it—he fights fraud for his job with the U.S. Securities and Exchange Commission. I had tearfully implored in my not-so-perfect Portuguese, "No, this doesn't make any sense. We have followed all of the rules, and there is no way we should have to pay two thousand dollars to right something that we never did wrong."

Amid it all, Max, Ella, and Maya had run and crawled about, attempting to amuse themselves without toys. They played on the filthy floor of the police station, they went out into the hot vacant courtyard in front, they tripped about on the uneven pavement with dry weeds poking through, and they

Figure 7 Maya and Ella moments before Maya learned to walk in the federal police station.

begged for food and water, not understanding that we couldn't leave our post lest we lose our place in the process and have to start all over again with our feeble, ineffective negotiations. Tired, we had hung out inside the station, as close as we could to the officers on duty, hoping to be more visible this way, to be noticed by someone, to be dealt with, and, in turn, to be able to leave. Suddenly, chubby, bald Maya stood up off the chair her sister was pushing her in and there, in the Ilhéus police station, took a step, then another. Her first steps in life. She looked up at us, at her brother and sister, and smiled a toothless grin in the sheer delight of a child doing something new. Despite the tensions of the day and having to pay a ridiculous fine for something we didn't do, in an instant our world was righted by the moments that matter.

Beyond the police station and the bustling public bus terminal adjacent to it is another place I always informally monitor, a stand of tarp-covered plastic structures that perch precariously on the edge of a mangrove. Sometimes a tattered flag hangs from one of what are makeshift homes, marking an identity.[2] I always think how no one should have to reside in a narrow strip between a busy highway and a mangrove, washing clothes in stagnant water and battling dengue-ridden mosquitoes that cohabit this uninhabitable space. But this is a reality of life here.

Past the small, makeshift community of creekside tents, I travel the winding road that leads out of Ilhéus, away from the coast, and inland. It is lined with gardening stores and cheap motels bearing skanky names in poor English—Lo-bi-Do, Paradise Palace, Cum-on-In. After passing these sites, I see a newly constructed housing site for the popular Minha Casa Minha Vida program, a remnant from former President Lula's Workers' Party era. It rises high along the side of the road.[3] Several gas stations offer fresh corn porridge for a snack, advertising the delicacy with enormous signs perched along the roadside should one happen to have such a hankering while speeding along the regional highway.

As I get closer to the university, I near the small villa called Banco da Vitória. At the entrance I look to my right. Dense Atlantic Forest looms high above the rocky cliffside. There is usually someone filling a large plastic jug of water from a stream spurting out into the roadside from a rusty-looking pipe as cars speed dangerously close to the narrow shoulder. Turning away from the hillside and paying attention to the precarious curves, I look to the other side of the highway to monitor the water level on the Cachoeira River. The river seems to look drier every year, but perhaps it is just my imagination. Coming into town, I slow for the three small speed bumps in succession. Teenagers hover at each, knowing that cars will slow down and that this is their chance to sell their wares of the day: bags of oranges, large cupuaçu

---

2. For formal groups engaged in land occupations, which can be both rural and urban, one of the first acts of occupation is to fly a flag marking the group's identity. People without land sometimes camp in visible but uninhabitable places like this mangrove in the hope that Brazil's federal agency in charge of land distribution will take notice. Campbell's comprehensive study of how people occupy, settle, colonize, and bring property to life to territories (like the Amazon) tells of settlers who hope to be awarded land through the Brazilian state by finding ways "to bring property to life" (2015, 5). This is what these settlers are trying to do on the edge of the mangrove.

3. Minha Casa, Minha Vida, or My House, My Life, is a nationwide public housing program that was introduced in 2009 under Brazil's progressive Workers' Party, called the Partido dos Trabalhadores (PT).

fruits, neatly cut and packaged jackfruit, strings of fish and crabs. I know I am getting close to my destination now.

There is one last place along this road I always monitor. For years it has been my regular stopping spot en route to UESC. I imagine that most people speed by, barely noticing this place, but here live friends of mine, people I know quite well. They pay attention to the traffic going by, and if they notice my fuscinha speed by without a stop to visit, the next time I see them they will ask why I didn't stop. I have spent the night here, tucked under a carefully laundered mosquito net people have removed from their own bed for me. I am always treated with respect and hospitality, for we trust, and care for, each other.

On this day, however, I watch some of my friends making their way across a field of grass toward their home on the hill next to the makeshift school, the fruit-drying factory, the cacao storage shed and bagaças. I slow to watch them, for I cannot help myself. I hope they don't look up. I long to call out, to stop my car, to go in and chat over *chocolate caseira*, homemade cacao squares, which I know they will have on hand in a large glass jar with a green plastic top that sits in the center of their kitchen table. But then I think about my family, the children in particular. I think about the promise I made to Jeff to stay away from here. He is worried about my safety. I continue on down the road past a simple sign marking this place I know well, the "Frei Vantuy Agro-Ecological Land Reform Settlement."

Meeting Maisa

Las Vegas is not my kind of town. The noise, the chaos, the neon, the vapid pursuit of a fix of all types. But when the Latin American Studies Association held its annual meeting there, drawing Latin Americans and Latin Americanists (as people who study Latin America are called) to board planes and dutifully converge to discuss history, culture, politics, economics, and literature in the company of slot machines, I went to Vegas. I, however, went for another reason: I went to meet a professor named Salvador

Salvador hails from Southern Brazil. He has light skin and thin, graying hair. He is calm and reserved but quietly friendly and reflective at the same time. He is also kind and well intentioned, the type of generous person who willingly meets with a random graduate student early in her career. I have a hard time imagining Salvador living in Wisconsin, the largely rural land of cheese and beer that is my home state, where he earned his doctorate in sociology long ago. Amid the ringing of the gambling apparatuses just below the conference rooms that day, Salvador and I sipped coffee and chatted in a mix of Portuguese and English. I hung on to his every word, impression, and piece of advice, for he was my introduction to parts unknown. Salvador lives and works in Southern Bahia but traveled to Las Vegas to give a paper on one of the key issues I wanted to learn about—agrarian reform.[4]

Years after our initial Las Vegas meeting, I find myself in Salvador's office during my first few weeks living in Bahia. It has been a Herculean effort to get here. My morning started badly, and I am exhausted though it is only nine

---

4. Land reform, also known as agrarian reform, takes place within the context of Brazil's Federal Constitution (revised in 1988), which states that unproductive properties are legally available for state expropriation and subsequent reallocation to families in need of land (Sauer 2006, 178; Wright and Wolford 2003). The National Institute of Colonization and Agrarian Reform (INCRA), the federal land reform agency, then awards this land to people organized into groups or associations. INCRA has the ultimate decision-making authority on how a farm can be divided and how many families each farm can support once it is expropriated. To create formal land reform settlements, INCRA determines the average land size for a family; these are called fiscal modules. These vary by state and are defined as the minimum amount of land for families to have enough to survive and thrive (Oliveira et al. 2020). Fiscal modules range in area from 5 to 110 hectares (ha) throughout Brazil and, in Bahia, from 7 to 70 ha. In the Ilhéus and Itacaré regions, the fiscal module is 20 ha (Landau et al. 2012, 47).

The push for agrarian reform varies by region throughout Brazil and is largely led by grassroots efforts (Branford and Rocha 2002; Cehelsky 2019; Ondetti 2008, 2010; Sauer 2006; Wolford 2004, 2005, 2006, 2010b). With the cacao crisis, a staggering number of approximately 200,000 rural workers lost their jobs on cacao plantations in the early 1990s (Demeter 1996). One team of researchers reports, "Although out-migration from the region was significant, unemployed rural workers constituted a reserve army of the poor for the organized land-reform movement" (Cullen, Alger, and Rambaldi 2005, 752). Wilder Robles's review of land reform in Brazil concludes, "After more than three decades of agrarian reform, Brazil remains a country with highly skewed landownership. Peasant-led agrarian reform efforts have had limited impact in changing this situation. Agrarian reform remains an unfulfilled political promise, and this situation continues to create tensions and conflicts in the countryside" (2018, 1). Others concur, noting that Brazil's land distribution is one of the most unequal in the world (Escallón 2019; Wittman 2009). Despite the constitutional provision, holding unproductive land is still financially attractive for tax purposes—government subsidies award low tax rates to large landowners, and land itself is a tax shelter for high inflation (Wolford 2001, 305). Given these structural incentives for large landowners, wide-scale agrarian reform has never been successfully implemented across Brazil as it has in other countries, like Mexico and Bolivia.

a.m. The septic system at our house is under repair, and to come and go we must walk across boards leading from the house to the street. The boards are perched precariously over an open sewage pit. Each morning it is a trying effort, at best, to keep Maya in the house while I transport Max and Ella safely across the hole on two-by-four planks. Failure to navigate this successfully means disaster. Today was particularly trying, as it was raining, the kids were crabby, and the planks were slippery. But we made it across the sewage abyss. Maya was at home with Guiomar, the two other children had been dropped at school, and I now had a coveted daylong window for research.

The air conditioner hums full force as Salvador and I chat casually about family farmers, the agrarian—or land reform—movement, and the specifics of these topics in Bahia. Decades ago, the region was engulfed in what people call *o grito do cacao*, the cry of cacao, which ravaged the local economy and recast social relations as former wage laborers scrambled for ways to obtain and farm their own land. But the fall of cacao has also brought something positive from the perspective of some. Various land reform organizations have arisen throughout the region. Garotinho, an employee of a national land reform group called the Movimento Luta Pela Terra (MLT), the Fight for the Land Movement, describes the origins of his group's involvement in Southern Bahia. "The need for the MLT came from *um grito no campo*, a cry from the fields," he says. "No one spoke about the problems here. No one said things are wrong here, and we need to do something. It was very difficult to speak of social movements. The MLT was the voice of the proletariat. . . . With many people out of work, there arose a cry. The *fazendeiros*, the large farmers, threw people off the land, and there weren't options for work. There were so few with a lot of land, and so many without any land at all." But as cacao colonels abandoned farms infected with the region-wide witches' broom disease, this opened up the possibility for those who needed land to band together and collectively settle tracts of property that had previously been owned by a single person.[5] Salvador and I chat about this history, about

---

5. See Trevisan (2019) for a deep analysis of how the cacao crisis led to increased land reform in Southern Bahia. The expropriation process can be long—it often takes years. As previously noted, for INCRA to award land to a group of people, it must be through an association— either through a formal, organized agrarian reform group or an association of people formed explicitly for the purpose of obtaining the land. In this latter case, a diverse group of people can opportunistically come together around the potential for obtaining land, rather than around a common history or ideology. Furthermore, processes of obtaining access to INCRA can be

this scramble for land and livelihood. As we talk, suddenly he pauses and cocks his head as if he has just thought of something pertinent. "Hey, if you want," he says, "I can take you to a land reform settlement right now—there's one nearby that I'm due to visit. We can see if folks are around." I nod eagerly.

Ten minutes later, we get out of Salvador's car and walk across a field to meet a small group of people gathered in a one-room school, housed below one of five bagaças.[6] We walk into a room filled with people squeezed into child-sized desks. They are clearly in the middle of a meeting. Messages of hope and inspiration printed in neat bubble letters on colorful poster paper adorn the cracked walls. Sunlight filters through dirty windows, which reveal golden cacao pods hanging on a nearby tree just beyond the door.

Apparently we have caught people after a long day of work. They are grubby but engaged in the meeting we have crashed. At a pause in the agenda, Salvador says a few words to explain our presence and then, without notice, gives me the floor to introduce myself. Those gathered listen politely to my stammering explanation of who I am and why I am here—a speech that is still awkward for me to give in Portuguese. Then, in turn, the people introduce themselves. One woman, Marlene, is the president of the land reform settlement. Genilse is the treasurer, and Henny serves as the settlement's secretary. As a fourth woman introduces herself as Maisa, the vice president, I look about and utter the first thing that comes into my head. "Geez, there sure are a lot of strong and powerful women leaders here!"[7] People chuckle and nod in agreement.

---

laden with personalism, relationships, histories, and interests and, at the same time, family farmers not affiliated with an agrarian reform group complain that a majority of INCRA funding and attention goes to formal agrarian reform groups, which also have hierarchies of power among them. Some here claim that NGOs in the region have added the term *social movement* in their names as a strategic ploy for power with INCRA (for the complex relationship between INCRA and social movements, see Wolford 2010a).

6. As Rui explained during a visit to this land reform settlement, "You can tell how big a cacao farm was based on the number of bagaças." He looked around the settlement and remarked, "and this was obviously a very productive farm with five bagaças."

7. While many grassroots activists are women, poverty in Brazil has been characterized as "feminine, landless and common" (Onsrud, Nichols, and Paixao 2006, i). Chandra Talpade Mohanty (2003, 64) argues that feminist struggles are waged in discursive as well as material ways, and the struggles I witnessed in Southern Bahia often brought the discursive and the material together in the practices of women farmers. Women also have a history of activism around land rights and sustainable development, particularly in the Brazilian Amazon with the rubber tappers movement (Harper and Rajan 2007, 336).

The story of how this community came to be is representative of the political, personal, and practical challenges faced by family farmers living on land settlements throughout the region, as well as of how the crisis of cacao created an opening for agrarian reform here. The former owner of the cacao farm, where the settlement now exists, was a Portuguese man. Although he wanted to will the farm to his daughter or son, they were not interested in a failing farm and already had careers in nearby cities. Given this situation and after suffering a stroke, the farmer signed the land over to the National Institute for Colonization and Agrarian Reform (INCRA), and it was placed in the government lineup of properties for agrarian reform. The Portuguese farmer died soon after the transfer occurred. The process of transferring the land from INCRA to a formal agrarian reform community was facilitated with the help of a nearby Catholic priest who was connected to a Catholic agrarian reform group, the Pastoral Land Commission.[8]

Recounting this history, the families who came to live in this community make it clear that they are not linked to any long-standing regional agrarian reform activist or to any other formal group. "To access INCRA," they note in the group meeting, "you must either be with a social movement like the MST [the Landless Rural Workers Movement, an agrarian reform movement] or you have to be an association." But some note that "the language and even the thinking are different" with these organized land reform movements. As the leaders of Frei Vantuy describe it, a group of people came together, growing through word of mouth and personal relations, in the hope that the cacao farm would eventually be theirs. They formed an association "to have more power, to be organized against an invasion by formal groups," Henny, a group leader, explains. "Most of the government's money is directed at social movements," Maisa chimes in, referring to groups like the MST, the MLT, and the MLST—all formal land reform groups in Bahia.[9] "But *we* weren't affiliated

8. The Pastoral Land Commission was founded in the Amazon region in 1975 and has grown to be one of the leading forces for land reform and the rights of the family farmer in Brazil today.

9. There are a host of land reform groups, also called movements, both in and of themselves and in relationship with each other. The Landless Rural Workers Movement, or Movimento dos Trabalhadores Rurais Sem Terra (MST), is Latin America's most well-known land reform movement. In Bahia, the Land Liberation Movement, or the Movimento de Libertação dos Sem Terra (MLST), was an agrarian reform group that later split into two groups, the MST and the Fight for the Land or Movimento de Luta Pela Terra (MLT). See DeVore's article, aptly titled "The Landless Invading the Landless" (2015), which reveals the complex relations between local land reform groups in Bahia and more established groups like the MST. Wolford (2010b)

with these movements," she explains, speaking of the families living in Frei Vantuy today. "Many people here set up an encampment on the side of the road, and some camped there for an entire year in the hopes of raising the visibility of their cause with INCRA," she says, motioning about the room. "They built tents, raised visibility, waited—even in the rain." Those around Maisa nod in agreement, murmuring in low voices, "*Pode creer*," you better believe it, in support of her account.

Another leader jumps in to help with the story, for it is not one person's tale but a collective history that belongs to all. "But just when it looked like INCRA was going to award us land," she says, "people from the MLT tried to get the property by arguing that because they were already an established organization they were more deserving of this property." I can almost feel the group getting agitated with this memory. I imagine a ragtag group camped for years and a more organized movement coming in, trying to usurp them. Someone else jumps into the conversation, indignantly stating, "But *we* were the ones waiting for this opportunity. We had been waiting for years. These other groups came in only once it looked like the land was really going to be turned into a formal agrarian reform settlement by INCRA!" Another person finishes the story, calling loudly from the back of the room, "But we persisted, we won, and here we are." People erupt in loud claps. They nod. They smile. The fight to get the land formally designated as a settlement took more than two years, and finally INCRA made the decision that the land could support thirty-nine families. The settlement was founded in 2000 and named after the Catholic priest who helped the group, Father, or Frei, Vantuy.

As the leaders and residents tell their story, I perceive that although she is the vice president, Maisa is the group's natural leader, the one most people look to for guidance.[10] She is light-skinned and dark-haired, with Italian blood and Bahian roots. Though I don't know it on this day, over time she will become one of my best friends in the region. I will come to know her as a devoted settlement leader and activist, as the mother of four children, as a farmer, as a businesswoman, as a strong, unwavering, and empathetic

---

also analyzes the complex dynamics between movements like the MST, which play a significant role in securing land, but then, once people are settled, tensions can arise around specific forms of agricultural practices, and Caldeira (2009) reveals how women are marginalized within the MST movement.

10. Maisa has served as president several times (2005–2006, 2008–2011, 2014–2017, and 2019–2021).

feminist leader. I will also come to know her as the settlement's most effective speaker, political lobbyist, strategist, advocate, and all-around champion who strives to provide all who live here with new opportunities that showcase the settlement's agricultural potential. When she pays attention—which is almost always—Maisa's eyes are piercing and bright. She watches everything and is quick to smile. And when she has something to contribute to the conversation, she speaks gently but firmly, in a voice that infers years of *a luta*, "the fight," that she is engaged in. But on this day, I am meeting Maisa for the first time. I have no idea of all that will be, of these ways I will come to know her. I also have no idea of all she will go through in the course of time, a fate that, strangely, Jeff seems to have foreseen.

Meeting Maria do Carmo

Months later, I head to a Ministry of Agricultural Development event near the city of Itabuna, one of the region's largest metropolitan areas, located thirty kilometers inland from Ilhéus. The event has the celebratory tenor of a political rally. Held mostly outdoors, it is meant to introduce family farmers and agrarian reform settlers to the latest governmental program of dividing Brazilian states into "territories." Participants have been bused in from around the state, most donning T-shirts they have been given to promote the territoriality concept that is printed boldly on the front.[11] Hour after hour, dignitaries line up at a head table to give speeches on the positive aspects of this new government-sponsored initiative. I learn that this is a typical style for public events in Brazil. Important people give rousing speeches while others file in and out of the official meeting area at will, holding side conversations in the back of the room and getting water or snacks as they need. Most of the farmers, however, sit quietly in their chairs and listen.

Outside the public speech area, booths are set up where vendors sell fresh produce, handicrafts, jewelry. I wander about, slightly bored by political speeches and in need of diversion. One booth in particular stands out. I stop to chat and compliment the two women staffing it, declaring they have

---

11. Starting in 2007, the Bahia state government adopted "Territories of Identity," a geopolitical concept that arose from the social movements linked to family farming and agrarian reform that was later adopted by the Ministry of Agrarian Development. Today there are twenty-seven territorial regions in Bahia.

"clearly the best booth of the fair." They chuckle and modestly inform me that they are also surprised by the attention they are getting. This is their first public fair. "Feel free to hang out here," they say, "to get to know our products." And so I do.

The women, Maria do Carmo, often called just Carmo, and Graça, are obviously sisters. Both are slight in build, with thick, short gray hair and tanned, unlined faces that defy their age. Both have twinkling eyes that are serious and intense, yet playful and welcoming. Both converse and laugh and joke with ease, welcoming all who stop by. The women have done an impressive job of making their simple structure inviting and eye-catching. Chocolate caseira, cacao liquor, and *doces*, sweets and jellies of mashed banana and cacao, are displayed on a starched red-and-white-checked tablecloth. What I particularly like is that sampling the products is not only encouraged but insisted upon. "Come and try," Carmo and Graça call out with an air of entrepreneurship to other passersby. I continue to hunker in, eagerly tasting all of the delicacies. In time, however, the event organizers come to round up those of us milling about the booths. "The afternoon program is beginning," they note, nodding toward the door. "Please head inside."

Again, the room is packed for hours of additional dignitary talks and dry NGO presentations, which focus on the Atlantic Forest flora and fauna while ironically neglecting to mention the human element of this region that integrally shapes conservation here. As I had learned long ago in the offices of WWF, modest-sized properties, or family farms, often lie within and adjacent to the Atlantic Forest. I know these speeches provide prime fieldwork material for me, but I struggle to follow it all with the loud microphone, the hot room, and the droning oratories and PowerPoint presentations in a foreign language that still demands my unwavering attention for comprehension. As the official talks near a close, the floor opens for those who are moved to provide their impromptu commentary on the day's topic—the process of organizing family agriculture into territories of production spread throughout the state. One by one, folks neatly line up behind the microphone, patiently waiting for their moment to speak. One woman passionately represents the perspective of Indigenous peoples living near Bahia's southernmost border. A man recites a poem on the natural and cultural characteristics of the region. Then, as the open forum is drawing to a close, I look up to see one of the sisters I met at the booth earlier take the podium. She stands as tall as she can, looks up at the packed auditorium, and begins. Her words are initially shaky, stilted, but quickly she seems to find her voice and purpose.

"My name is Carmo. I am the daughter of a family farmer. Though I was living in São Paulo for a time, I left the city ten years ago and came back to this region, to *my* region, to invest in a modest-sized property here." Carmo pauses, looks around, and then continues. "And since that day, a day long ago, I have been working every day, *every* day," she emphasizes. She holds up her hands for all to see. "I have been working with these very hands to make a life for myself." People halt their side conversations, and the room quiets noticeably.

All eyes turn to Carmo as she continues. "I came up here to speak to you all today because it became clear to me as I was listening that what we have here is a 'complete table.'" She gestures to a table in the front of the room adorned with regional fruits and vegetables. But Carmo's words have a deeper, metaphorical meaning beyond the obvious table in front of us all. She gestures toward the room. "I'm not sure if you have observed, but we have people from the land, *a terra*, all the way to Brasília right here in this room, here in front of us."[12] She looks around at the farmers in the audience and the dignitaries in front. "And it is important that we take advantage of this," she continues, "for these are opportunities we have in life, unique moments. I mean, if the same group of people comes together tomorrow, it isn't going to be the same as today. This is a dance of life, and we are all present. So I don't know how I can manage to be here and not participate, not express my sentiments." Carmo pauses, dramatically, and again looks around, her voice growing stronger, gaining confidence now that she has captured the attention of all. "I don't speak as a single person; I speak for *all* the people who came with me here. What we perceive, as family farmers, is that the current structure doesn't elevate us. It doesn't help us." I catch a few dignitaries giving sidelong glances at each other. They seem to be wondering where she is going with this commentary at what is meant to be a celebratory kickoff event for the state's program. Carmo continues on for a good twenty minutes, using captivating words, metaphors, examples, and stories from her daily life that command people's attention. Many of the farmers in the room nod their heads in agreement while some of the governmental leaders continue to shift in their seats or stare straight ahead, expressionless. She ends with a rousing finale. "And by the way, people, I hate the word *small*! 'Small producer,' 'small worker.' Small?" she

---

12. "All the way to Brasília" indicates from the municipal, regional, or state level up to the federal governmental level, which is run from the Federal District of Brasília.

says, emphatically. "Enough of small! If we, with sweat on our faces, can bring food to the whole world, why in God's name are we called small?"[13] A tear rolls down her face. She wipes it away and continues as if it hasn't happened. I, too, cannot contain myself, and I wipe the tears from my eyes while hoping that no one notices an outsider, a gringa, who is strangely emotional at this political event.

Back in Itacaré that night, I chat with my friend Catu, a nativo guy and local community leader. We sit on his front step, watching people walk by on their evening stroll about town. The night is dark, yet busy. Street vendors sell bean cakes, popcorn, ice cream. Kids run about the town square. Still excited about the day, I tell him about meeting Carmo and Graça. As always, Catu listens thoughtfully and then offers his own commentary on the situation as I describe it to him. He has an opinion on how hard it is to actually make ends meet here as a farmer.

"A majority of farmers in Itacaré have land but still have to work for large farmers, fazendeiros, you know. Their wives are the ones who work in their gardens, so you have so-called family farmers working for large farmers." He continues, emphasizing his point. "I went to visit a farming family that I know in the countryside, and they invited me into their home. And you know what? I looked around and noticed that the *forno da lenha*, the open-air wood-burning stove, had cobwebs in it." He pauses, looks at me, then repeats for emphasis, "Cobwebs! Do you know what this means? Do you know?" I shake my head no, not quite getting his point. "This means that the family, who were *farmers*, hadn't had food to cook for several days! They were subsisting on coconuts and *farinha*, or manioc flour, and probably working for someone else!"[14] Catu states this, matter-of-factly, shaking his head in dismay. "Yeah, the plight of the family farmer here is rough indeed." He sighs and looks away. As we chat, I, too, look out at the people strolling by. These people are decidedly not family farmers. They shop at the weekly fresh farmers' market, at supermarkets. The farmers in this region live in

13. In an analysis of Brazil's agricultural modernization process, Guanziroli et al. note that family farmers, until the mid-1990s, were described in various ways—as smallholders, small producers, subsistence farmers, huskers, settlers, peasants, or low-income farmers (2019, 176). But linguistic traditions die hard, and to this day "small farmers" is still often used.

14. Though less common, other farmers have success stories. They have managed to join with NGOs or other groups of farmers to access credit, and they are making their living on the land in ways that challenge the historical composition of agriculture that, for the past 200 years, has been dominated by the latifúndio.

the countryside; they are home in bed by now. Like Carmo, they work every day. I think about how family farmers are going hungry in a region rich with agricultural potential. And I also think about how I will never again use the phrase "small farmer." Nothing about these people's realities is small.

Perceiving the "Other"

It is a hot afternoon in Bahia. The air doesn't move. Time, too, stands still. It seems the entire region is lazing in hammocks after lunch, digesting, swaying, waiting for the sun to drop a bit and the heat to subside. Carla and I sit on the porch of her upper-middle-class home chatting amiably. A blond-haired woman of European descent, Carla is the daughter of a former large landowner and is now in charge of projects working with farmers throughout the region that are managed by the NGO she works for. We sit in lounge chairs while the woman who works in her home brings us tea. Our conversation shifts from family and friends to farms and land reform settlements, some of which are being established on the edge of the region's remaining Atlantic Forest. Carla looks at me and says, with a somewhat conspiratorial laugh, "You know, *neguinho* will invade and rob wood when they can.[15] You need to enclose everything. You need to patrol the area." The conservation NGO she is working with has a project with farmers; the reason it does so is to prevent incursion on the region's forest.

Carla continues, carefully explaining how deforestation in the region is a result of poverty, a fact that is not altogether untrue. The environmental-agrarian tension that Carla talks about is rooted in Southern Bahia's more recent history. When agrarian reform settlements sprouted up after the cacao crisis hit in the late 1980s, these settlements were often established on former cacao farms that contained Atlantic Forest. Furthermore, deforestation in Southern Bahia was higher in the early 1990s than it was in the 1980s, when cacao was still king.[16] This reality began to create a perception of farmers associated with land reform as destructive, pitting environmentalists who

---

15. *Neguinho* is the slang term for "guy" that can be used in informal conversation; it means "little dark one." However, when the person uttering this phrase is a white upper-class woman (as was this situation), the term seems racist and derogatory (see Roth-Gordon 2007, 2016). As Goldstein observes, "The middle and upper classes need this 'low Other' in order to know who they are" (2003, 86).

16. Cullen, Alger, and Rambaldi (2005, 752).

were trying to conserve the remaining Atlantic Forest against land reform settlers who were trying to eke out a living on small plots of land or former cacao farms. As I had perceived long ago in the offices of the WWF, and as environmentalist friends sometimes indicated, environmentalists and farmers don't always align. When speaking about conservation challenges in the Atlantic Forest, one friend had emotionally exclaimed to me, "I hate agrarian reform! It is a thorn in my side."[17]

As I often found, there were more behind strong statements like this. In Carla's opinion, there are deeper cultural differences. "These folks simply 'lack civilization,'" she confesses to me, speaking of farmers connected to land reform movements. I ask her, "What does that mean exactly? What does it mean to 'lack civilization'?" I look at her quizzically. "Well," she continues, "when you have a land reform settlement, you have to put in water, a bathroom. And sometimes they don't do this. Sometimes I go to land reform settlements and go the whole day without going to the bathroom because I'm not going to go in the forest!" Carla says, implying that inadequate bathroom services is a choice, rather than a result of an inability to connect to a municipal sewage system or lack of funding to even make the bathroom structure.[18]

A month later, I am on the land reform settlement with Maisa. I chuckle to myself, thinking of Carla's judgments; things here are nothing like she had described. As we stroll from her house to her cacao plot, Maisa and I look for the easiest route to pass in shadows of the tall forest trees, which shade us from the scorching sun. We step slowly, carefully, our feet crunching fallen leaves on the forest floor. Maisa taps into subjects that are often initially inexplicable, nuanced, hidden from view, like the cacao under the canopy. We chat about land reform, and Maisa carefully explains how she sees it. "With agrarian reform," she says, "to the general public we are seen as vagabonds, as people who don't want to do any work, as robbers who invade, wanting to

---

17. Environmental NGO staff sometimes work with large landowners (with more than 50 hectares) to form partnerships that set aside some of their land as private reserves, or RPPNs. As many large farms in the region had intact forested areas that weren't suitable for farming yet had high conservation value, RPPNs became an important conservation strategy. However, while having an RPPN positions the medium- or large-scale farmer in a positive, altruistic category, the family farmer more often continues to be seen as the uneducated anti-environmentalist—particularly when he or she lives on a land reform settlement.

18. See Wolford (2005) for a detailed analysis of the clash between elite perceptions of landlessness particularly tied to the MST in Southern Brazil.

steal." She pauses, steps carefully around a pile of cacao on the forest floor that is waiting to be collected and broken open to dry on the bagaças nearby, and then continues her reflection. "This is a little old for us. All that has been given to us we've had to pay for, we've had to develop. We don't get anything for free or given to us. No, we don't. And if the worker doesn't go to the field to work, to farm, to fight with their own sweat to achieve what is theirs, they will have to leave. They can't stay here anymore." She pauses, talking about the community where she lives. She reaches down to pick up a cacao pod, breaks it open, and gives it to me to extract the seeds, suck off the milky white pulp, and then spit them out. We continue walking, and she resumes her assessment of how they, as family farmers on a land reform settlement, are viewed. "When we're considered *sem terra*, landless," she says, "we're not really seen . . . or we end up being seen in a certain way. We aren't seen like the large farmers with their ample fields . . . we aren't viewed as producers; we don't have any value." She pauses and concludes with a phrase that makes no sense in English, but I inherently understand because of its obvious, visceral

Figure 8 Edvaldo drying cacao on a bagaça at Frei Vantuy.

meaning. Maisa sighs and simply states, "Attitudes like this kind of kick you in the shins." She continues leading me along the path.

Perceptions, imaginations, and judgments can converge into a singular and static stereotype, particularly when people think about the land reform movement in Brazil. This stereotype can even pit agrarian reform activists against one another. We arrive at Maisa's cacao plot and start to work, gathering up the fruit in baskets. The birds in the forest fill the silence between us. It is shady, cool even, under the canopy of the trees. Maisa seems to still be thinking of our conversation, and she offers, "Agrarian reform can elevate people. It is a way, you know, for a human being to change their life. You think when you come to take the land your life is going to change, there is going to be an effect from this. You think, 'I'm going to have my work; from here I'm going to develop and my family is going to grow.'"

I look at her, then back out at the forest that envelops us. There is something I can't voice but also can't stop thinking about. I can't tell Maisa about an interaction I had a few months earlier in the chic apartment of one of Brazil's foremost social scientists, an internationally renowned scholar. I liked the guy—he had spent a stint teaching at an elite university in the United States, and as my sister was his student during this time, we had received a generous invitation to join him for Sunday lunch at his home, one of the nicest offers of hospitality one can receive in Brazil. This scholar had authored dozens of books on Brazilian society, culture, and politics and was a brilliant, kind, affable man. After a veritable feast, the professor brought out a bottle of whiskey, and the conversation turned to my work. Though at the time I had yet to begin my research in earnest, I explained my interest in looking at the interactions between the environmental and land reform movements in Brazil, which despite often being at odds, were also interdependent given the proximity of forests and farms. Listening to me, the professor had leaned back in his chair, laughed, and, stretching his arms behind his head, rhetorically exclaimed, "Can you imagine those agrarian reform people coming into your house, sleeping in your bed, wearing your clothes? Can you imagine this? How terrible it would be!" He shook his head in dismay.[19] I remained

_____

19. Violence, as this scholar seems to imply, is neither condoned by the leadership of the formal agrarian reform groups nor a common practice (Hammond 2009; Rosset, Patel, and Courville 2006). One reason this stereotype exists is that the media is often present when protest marches or demonstrations for agrarian reform take place. While the media coverage is in part why activists employ public consciousness-raising tactics, it can also work to their disadvantage. The images that live on in people's minds are often dramatic ones—an activist

quiet, uncomfortable, unsure of how to counter him as I was a neophyte in his country. Looking at Maisa now, I felt guilty for my silence.

As I soon learn, the scholar's perception is not an isolated one. A few months later, Maisa and I are seated in the stifling hot waiting room of the secretary of education's office in Ilhéus, located among neat rows of upper-class homes. We have a gorgeous view of the palm trees swaying in the breeze, the Atlantic Ocean surf gently rolling in beyond. As people wait for their appointments with the secretary, they chat politely, making small talk and new acquaintances before they are called in for their brief audience. They wish each other good luck as they enter the closed-door office one by one. Maisa chats amiably with a stranger, as she often does, when a friend of hers enters the office. After kisses and a hug, her friend turns to her and asks how things are going at Frei Vantuy. Without a beat, the stranger Maisa has been conversing with turns with wide eyes and questions Maisa, incredulously, "*You're* landless?" Maisa and I catch each other's eye. We both know that she, a seemingly middle-class white woman with key political connections like the secretary of education, doesn't fit the picture of a land reform settler. We know the struggles she has endured to earn—and keep—a place in this community. However, we also know that being identified with a land reform settlement can saddle people with particular stereotypes and judgments that speak to class relations, an indelible reality of Southern Bahia.[20] Maisa turns to her new friend, smiles, and politely replies, "Yes, I live in the

---

raising his fist against the gun or stick of a police officer or people overturning a desk—isolated events that then occupy space on television stations around the country for a few days and live on in people's memories. In this way, particular memories are created and sustained while the images of a peaceful march remain hidden. As Walter Benjamin asks, do memories pause on events that conjure up danger (1969, 255)?

20. Having land or not, being part of a formal agrarian reform group or not, or farming for yourself or for others are often conflated into one category that denotes a distinct social class. According to Plummer and Ranum, "In Brazil, the poorest of the poor historically have been the 'sem-terra'—the landless, a nickname given to the social class of rural workers who work land without having title to it. They work as tenant farmers, agricultural workers on large fazendas (plantations) cultivating crops for export, or as migrant workers" (2000, n.p.). The identity of the MST, and the broader agrarian reform question that persists throughout Brazil, elicit strong reactions—both of the need to fight for rights and of the threat this group of impoverished people who are gaining consciousness about their oppression brings to the elite-skewed status quo. For this reason, when family farmers are assumed to be landless, even when they are not, they take on these (mis)perceptions of class despite, in many cases, their upwardly mobile status (see Wolford 2005) and are socially (mis)categorized as "landless" in Brazil—even if they actually own land. Impressions such as these have real effects. They can curtail farmers'

Frei Vantuy community," and patiently continues the conversation with her new acquaintance.

Later, I ask Maisa if the woman's comment bothers her, as I assume it does. She brushes it off. "These people don't know us, they don't know *me*," she emphasizes. "Instead, they think of us as violent, as lazy, as waiting for a handout from the government." I think about the professor, about how friends ask if I'll be "safe" working with settlers living in land reform communities during my fieldwork. They are unaware that violence most often occurs against agrarian reform activists, rather than by them. While there have been violent confrontations among large landowners, the latifúndio, and aspiring family farmers, in such confrontations the majority of those hurt and even killed are the landless and agrarian reform affiliates.[21]

At the same time, land reform settlers and family farmers are a diverse, heterogeneous group. Even among themselves, they can carry stereotypes of "the other." Maisa pauses, then continues, reflecting on the problem with singular images or stories that can arise from isolated incidents or pointed actions. In turn, through the media, they are circulated around the country, become etched in people's minds, and feed and perpetuate stereotypes, impressions, judgments. She states, "You know, the vision and the way that these movements are represented today on the television is horrible . . . the reports we see of land reform activists overthrowing trucks, you know, you understand? These types of things transform the image of everyone else who isn't at fault at all." She sighs and looks away.[22]

---

access to civil society and state structures such as agricultural assistance, technical training, and financing programs.

21. Trends like this continue, as with the murder of Sister Dorothy Stang in 2005, which captured worldwide media attention with the documentary *They Killed Sister Dorothy*. There is an extensive literature on land reform in Brazil. The Pastoral Land Commission and Amnesty International regularly report on violence in Brazil against land reform (and environmental) activists, a trend that is only worsening under the Bolsonaro administration, which has repeatedly attacked land demarcation efforts by both Indigenous peoples and land reform activists like the MST (see Foley 2019; Garcia 2019). Wolford's landmark ethnography demonstrates how land reform politics are "messy, uncertain, changeable, not always knowable" (2010b, 25).

22. As Maisa's perspective reveals, partial and often skewed representations of one agrarian reform group can color and prevent alliances across the diversity of groups working for agrarian reform and the rights of family farmers, almost through a "divide-and-conquer" tactic.

Running into the MST

On a rainy afternoon in Bahia, I hibernate at home, trying to write and save my computer from the mist that inevitably enters through the roof whenever it rains, a common occurrence in the rainforest. The smell of black beans and onion wafts up and distracts me. The kids are trying to amuse themselves as they, too, avoid the water droplets that find openings in the clay tiles, drip upon us, and pool in the lower areas of the cement floor of our home. Max and Ella play with their pet turtles, Barbie and Tom. Maya makes hungry squawking noises and toddles about now that she is walking. I know she is looking for food and is likely tugging at Guiomar's leg, wanting to be picked up while she attempts to cook. They are inseparable, Maya and Guiomar.

As I type, a young man assembles a wooden desk that I have purchased in an effort to impose order into the chaos that is my life here. It is a respectable desk, a kind of rustic-chic piece of furniture, though I wonder where the wood came from. I could not help but think of illegal timber harvesting in the state park when I bought it. As the guy pounds away with a rubber mallet, at one point he catches my eye. "Why are you in Brazil?" he questions. I have no interest in making small talk during my precious and limited work time, yet I am too polite to say this. I launch into a canned description about studying the relationship between environmental conservation and land reform movements and assume he will soon turn away in boredom and let me return to my work. After my brief explanation, however, he pauses his work, turns his attention solely to me, and asks a surprisingly pointed question. "Have you been able to speak with anyone from the MST?" I stammer something about how I've only recently arrived in the region and I'm still working on making contacts, wondering how this guy knows about the MST. I assure him that yes, I'd love to talk with someone who knows about this movement. Again, he looks up, "Well, I'm a militant in the MST," he says, using a term to describe activists in this movement. "My name is Carlos," he says, extending his hand as he stands and gathers up his tools to leave. "I'll take you to see some places if you want."

A week later I meet up with Carlos. He is waiting in the rain outside a gas station in Itabuna with MST literature in hand, protected from the rain in a plastic folder. He jumps into my car to make our way to the organization's regional headquarters, directing me through streets submerged in water. As we park and run through the downpour, I expect to see a big sign marking the offices of the country's most well-known agrarian reform group, but instead Carlos pauses near an unmarked door and rings an intercom. After we are identified, we are buzzed in. We walk up a narrow staircase where, again, we are buzzed in through a secured door. This is not lost on me; I think about the violence against these groups and how I was naïve to assume the MST would have an open office.

We enter a small, sparsely furnished room where a handful of MST activists are meeting. They are focused and barely look up at us. April 17 is quickly approaching, marking the anniversary of the most violent day in the history of the MST when nineteen militants were massacred in cold blood.[23] The activists are preparing for a consciousness-raising march that will involve meeting in a nearby city and walking the final 100 kilometers to Bahia's capital of Salvador. I glance down at a large adjoining garage space with piles of food on the floor—beans, rice, powdered milk, tapioca, and such. "That's for the march, as well as for families in the MST settlements," Carlos explains, anticipating my question. "You should come with us," he invites me. I pause, thinking seriously about this. This is exactly what I want to do. In seconds, I ponder the possibility of leaving Max and Ella, who could never walk so far, and strapping Maya, who is still nursing, to my chest for the 100-kilometer journey. I believe, for an instant, that this is actually a viable option. We can camp; it will be great. I mentally calculate the days this will take. And then I stop. As much as I want to, I know I cannot leave Max and Ella while I hike 100 kilometers carrying a toddler. I look at Carlos, and simply say, "I would love to. But I can't."

A few weeks later Carlos and I meet again, this time in a small, noisy restaurant. I want to learn more about how he came to be active in the MST movement. Over cups of lemonade while the rain predictably torrents outside, he patiently explains that "the MST stems from 'the fight,'" a luta, as they say in Portuguese. Carlos elaborates for clarity: "Not hand-to-hand combat or violent resistance but a revolutionary perspective that results when, over time, a group of people are very persecuted." I nod as I follow him, thinking of Maisa's negative impressions of the MST. "This fight and people's affiliation

---

23. This march was to happen in April, a key month for the MST to raise consciousness about the plight of agrarian reform due to the 1996 El Dorado do Carajás massacre, when military police fired upon agrarian reform activists, killing nineteen and injuring many more.

with the MST," he continues, "is a last resort for people to create a life for themselves." I think about how these efforts to create a life, and how Brazil's constitution, one of the most progressive in the Western Hemisphere, technically allows for land reform to take place. But while land reform is possible theoretically, in practice many people are denied real opportunities.[24] "So taking to the streets, making this constitutional demand visible, is part of a strategic process," he explains. The waitress appears as he says this, looks at him, and brings us more lemonade. She senses we'll be here for a while.

----

24. Carlos's reference to "the fight" is a common phrase in Bahia among both land reform and individual family farmers. When asked how they are doing, a usual response from folks here is *"Estou aqui, na luta,"* or "I'm here, in the fight." But while resistance and change to the status quo is the foundation of the agrarian reform movement throughout Brazil, diverse agrarian reform groups have largely held to the ideology that they are armed only with the tools of the farm—picks and rakes—and that their purpose is to raise awareness of their plight through peaceful tactics like marching, demonstrations, and occupations of unproductive land.

Nevertheless, the public perception of landless activists as violent and lazy persists. Agrarian reform settlers bemoan how the general public view them as living on the settlement in houses that were built with public funding. Farmers have to deal constantly with profound misperceptions about who they are and how they work, with others' stereotypes of them that communicate deeper conceptions of their honesty, reliability, peacefulness, and trustworthiness. As José, a farmer living on a land reform settlement in Bahia, once lamented to me, "Everyone thinks we're just lazy, that we sit around watching television in our government homes all day waiting for a government handout." Sociologist José de Souza Martins observes that the vastly diverse and heterogeneous rural poor are often collapsed into a uniform "Brazilian peasantry" (2002, 322). Martins also notes that the way middle-class mediators construct and conflate all farmers is interesting to consider in the context of Southern Bahia, where these mediators are most often state and civil society actors. Agrarian reform settlers and independent family farmers often face stereotypes that are applied to both groups, such as perceptions of poverty, violence, and anti-environmental behavior. In other words, the perspectives, situations, and realities of family farmers and land reform activists are similar, whether land is farmed as part of a land reform settlement or individually.

While experiencing similar stereotypes, land reform activists and farmers also often have relations that transcend pure identification with one group or another. Maisa is a family farmer living on a land reform settlement with no group affiliation. Carmo is an independent farmer who lives on her own land and works on her own; however, her sister Graça lives on a land reform settlement. While boundaries across these groups are often fluid, important distinctions also exist: family farmers living on agrarian reform settlements must confront myriad challenges of living in community and forging a collective struggle to make a better life together, while independent family farmers face other difficulties, such as gaining access to capital, establishing a good reputation "on their own," and locating technical assistance as well as a social safety net in difficult times.

Carlos and I chat for hours, covering his past and how he came to be where he is in the present. "I joined the MST when I was a teenager, walking with my parents in a public march for consciousness-raising," he says. "When we arrived at the end of the march, I had decided to join the movement." He looks at me intently, making sure I understand the significance of what he will say next. "So my parents went home, and the MST became my father." As if intuiting that I want to ask him why our serendipitous meeting happened in the first place, he offers, "I'm just working in Itacaré for a bit now, to make some money. As soon as I can, though, I'll join the movement again." He signals to the waitress that we are done here. We pay and stand to leave. As we give each other a quick hug outside the restaurant, Carlos turns his tall, lanky frame, ready to amble off in the light rain, the noise, the traffic, and chaos of this rough neighborhood in the city of Itabuna. Suddenly, he looks back at me intently as if he has one last thing to say, which he does. "Remember, land signifies power," he says, and nods so slightly it is almost imperceptible. He turns again, walks off, and quickly disappears into the city.

Producing Culture from the Margins

In time, the Frei Vantuy Agro-Ecological Land Reform Settlement has become *the* foremost example of a model agrarian reform settlement in Southern Bahia. When Salvador and I had first visited the community, I was struck by the noise and busyness of its location, which seemed the opposite of the bucolic, quiet farming community I had envisioned. But soon I learned that this perception was my own romantic notion. While people's plots for farming are set back from the highway, nestled beneath the Atlantic Forest, isolated farming communities are harder to sustain. Farmers need access to markets, and the location of Frei Vantuy is ideal; it allows both products and people to circulate.

Life is also organized here: it has to be. Thirty-nine families living communally require rules. Most families farm for their main source of income, though a handful of households work in Ilhéus. Each family has its own plot of land, and there are also communal work responsibilities; families must help with collective labor two days per month, a strict requirement that, if skirted, may result in eviction. The farm still produces cacao, but the lessons of the crisis have been learned. Families have diversified their production, growing bananas, jackfruit, mangoes, papaya, peppers, and a host of other crops.

Beyond crop diversification, Frei Vantuy has other progressive initiatives taking root that solidify its enviable reputation. The settlement has a new fruit-processing endeavor, which takes the fruit grown on the settlement, dries it, and delivers it to schools in the region, serving as payment in produce for the financing the farming families are awarded through a governmental program.[25] The fruit-drying endeavor also provides much-needed income that farmers depend upon to buy seeds, plants, and farming equipment.

But there is more. Each year the residents of Frei Vantuy seem to have something new afoot. One year there is a special sheep-raising project that is communally funded and managed by a group of twenty or so families. Another year there is an effort to make handicrafts and purses out of coconut shells. It seems each time I come, Frei Vantuy is reinventing itself in pursuit of the latest, most promising opportunity. Negotiating and winning contracts with municipal leaders, dealing with banks to secure lines of credit, joining together and organizing to make something happen—this is the stuff of innovative, successful family farmers who give a new name to land reform. One day, as Maisa gives an interview to a group of reporters, I hear her declare, "We were able to negotiate our debt, and today we are proud of our achievements." I know, however, that this wasn't easy and was due in large part to strong leadership.

Beyond this, there is more—sheer tenacity. If farmers want to gain formal access to land through the agrarian reform movement processes, they must work with INCRA, the federal agrarian reform agency based in Brasília with branches throughout the country. But a person cannot gain access to land through INCRA as an individual; alliances must be forged through an association.[26] And if you want to form an association, you must have an elected

---

25. Farmers' debt is repaid through the Brazilian National Supply Company (CONAB), which buys their produce.

26. The processes of *associativismo*, of creating associations, has been studied in Southern Bahia (Couto 2007). One of the most important aspects of such processes, however, is that they reveal farmers' attempts to "re-present" themselves within the structures that government and civil society actors deem important. Sauer also points out the following requirement for governmental programs: "landless families must organize an association (which a priori is a forum for participation and decision making) through which they can then apply for funding" (2006, 186). These structural realities have led to a blending of different movements. For example, environmental NGOs sometimes brand themselves social movements to access INCRA, or agrarian reform settlements identify themselves as environmental settlements, as in the case

president and vice president, a name, bylaws. You must become a modern-day social organization.[27] This process has created the impression that the spaces and structures of power are increasingly merging into *um orgão só*, as people sometimes describe the INCRA process, meaning a single, monolithic bureaucratic state organization that holds the power to grant land and credit and open the doors to the projects that family farmers need. But behind every governmental agency are people, a fact that is not lost on Maisa and others at Frei Vantuy. Plus, Brazilians are known for something called *jeitinhos*, crafty and creative ways of getting around the pervasive bureaucratic systems throughout the country that can, quite literally, occupy days, weeks, months, and years of people's lives.[28]

On yet another scorching day in Bahia, I find myself at yet another public meeting in a stark cement government building. One by one the public figures start to speak. Though I begin the meeting listening attentively, before long the addresses began to blend into each other, the air stagnates even more, and I find myself nodding off, snapped back to attention from time to time by the polite audience applause separating one speaker from another.

After the traditional opening ceremony, we break for lunch. The dignitaries are to be shuttled off in air-conditioned cars to a model agrarian reform settlement, where a typical rural lunch has been prepared. People scurry about, getting on buses, and suddenly, somehow, I realize that this "model settlement" is Frei Vantuy, of all places. Amid the hoopla and fanfare that mark Brazilian public meetings, Maisa has managed to secure a special visit from Bahia's state secretary of agriculture, luring him to her home turf to feast on the cuisine of one of the community's best cooks, Dona Lu. Without hesitation, I find a way to get myself an invitation, to watch this jeitinho in action.

We arrive at our lunch destination where *galinha caipira*, a delicacy of free-range chicken frequently advertised along roadsides throughout Bahia, is being prepared. Dona Lu's house is a simple structure on the edge of the busy regional highway. As always, it is impeccably clean. A neat assortment of plates, cups, and cutlery are stacked on the table, topped with a blue-checked plastic tablecloth and a vase of fresh flowers.

---

of Frei Vantuy (which is formally named the Frei Vantuy Agro-Ecological Agrarian Reform Settlement), to demonstrate their environmental orientation.

27. See Ondetti (2008, 2010, 2016).

28. See Hess and Da Matta (1995). This "system" with which people must engage is monstrously large and time-consuming. In fact, in Brazil there is even a profession called a *despachante*—a person paid to facilitate bureaucratic processes.

The visitors mill about, making small talk while waiting for the food to be served. When the guests look at their watches, I can tell that the residents of Frei Vantuy are growing nervous: they worry that the lunch is taking too long to be served and that perhaps the VIPs will get antsy and decide to leave. People shift nervously in their seats, swatting flies, clearing their throats, knowing they have only a limited time to showcase their situation. But when the secretary sits back in a hammock, looks out at the view, and remarks, "What a *moradia boa*, a good place to live!" they smile and relax a bit.

Finally, the food arrives. All present feast on chicken, rice, beans, and salad, and Dona Lu brings out her specialties at the end of the meal, *mel de cacao*, cacao liqueur, along with some chocolate caseira. There is a purpose in this meal—to show governmental officials that the "former landless" are respectable, organized, dignified, and people *de confiança*, trustworthy. Later, when the state officials are asked to allot funding to a project, they might remember Frei Vantuy. If the settlement needs approval for something, they may be regarded well. Perhaps they will even change someone's impression of a family farmer living on a land reform settlement.

Maisa never openly speaks of these strategies to gain access, to alter impressions, in seemingly unassuming ways. Yet through the simple but profound act of hosting a meal, of bringing a group of strangers into their homes, she is working ardently to dismantle barriers of class, culture, and power. While the stereotype of the landless, rogue peasant positions family farmers in certain ways, at the same time Maisa and others are conscious, not passive, subjects engaged in reworking their positions to gain more agency and new spaces of power in the shifting agricultural context of Southern Bahia today. While no words about specific projects are exchanged, hospitality and conversation are. The best-case scenario is that the visitors leave with a new perception of the identity of family farmers, and perhaps over time this can translate into some form of political, technical, or even financial support.[29] Ideally, this not

---

29. Subjectivity refers to human processes and human agency within the broader structures of economy, politics, violence, and social suffering (Biehl, Good, and Kleinman 2007, 1). The notion of subjectivity allows us to look at individual as well as collective agency, which is "a necessary part of understanding how people (try to) act on the world even as they are acted upon" (Ortner 2005, 34). Seeing family farmers through a lens of conscious subjectivity allows us to understand how histories and class positions fuse in particular ways to create family farmers as subjects and how farmers can, in turn, knowingly manipulate their subject positions (see Giddens 1979; Roseberry 1997). Looking at the transformation of supposedly "landless, rogue, and anti-environmental" farmers into competent, environmentally minded citizens of the world

Figure 9 Geraldo, Antônio, and Josadabs in front of cacao on the Frei Vantuy
Agro-Ecological Land Reform Settlement.

only enhances the farmers' ability to access economic, political, social, and
environmental power but also affords them more agency within these spaces.
Before people leave, Maisa encourages fellow farmers to pose for a photo with
the cacao they produce. They gaze out at the camera, smiling, representing.

Hiding the Knife and Boots

Nearly two decades ago, the interconnections between agriculture and environmental conservation were only beginning to take hold across Southern
Bahia. Now they are becoming mainstreamed. One community receives a
prize from the United Nations for having a biodigester, which turns plant
material into energy.[30] Other land reform settlements begin the process of
obtaining formal certification for organic crop production.[31] Land reform

---

also involves "teasing out, historically and ethnographically, the various ways in which room for
maneuver is present but never unconstrained" (Li 2000, 153).

30. A biodigester can save one ton of wood per month from being cut and can then be
used to dry the cacao produced on land reform settlements and to produce biofertilizer.

31. The connection to environmental practices also ties into farmers' reworking of their
subject positions. Arun Agrawal argues that the creation of and change in human subjectivi-

leaders speak of efforts to make their settlements more environmentally conscious, and everyone seems increasingly tuned into not only the philosophy but also the practices of environmentalism in a collective attempt to undo the "anti-environmentalist" notion of family farming.[32]

Independent family farmers, like Carmo, are also fervently reworking the image of family farmers. They share production technologies and secrets, form associations, and try to find their way both individually and collectively.[33] Carmo has her finger on the pulse of all that is happening in the region and processes the transitions taking place with a deft combination of philosophy and practicality.

Carmo and I develop a routine of meeting so she can patiently explain the latest in her world. One time we converge in her hometown at the residence of her elderly mother, a remarkable woman who birthed twenty-six children and—miraculously—lived to raise them. Another time, we visit the land reform settlement where her sister Graça lives. Other times, we meet in Ilhéus for lunch or Itabuna for ice cream. Once, Carmo and her niece Alba even travel six hours via bus to Itacaré, where I live, where we frolic on the beach, take pictures by the ocean, and feast on seafood. Our meetings are

---

ties is a product of environmental regulation and Foucauldian governmentality, a process he terms *environmentality* (2005, 8). Agrawal looks to political and institutional processes as the impetus for the creation of particular environmental subjectivities, which is useful for thinking through the strivings of family farmers in Bahia, as perhaps no group is more connected with the environment than those who make their living from it. There are multiple connections between farmers and government processes (even when those processes are administered by NGOs) through land acquisition, access to credit, environmental restrictions, and the like. However, subject formation can also be approached more broadly. We need to pay attention to the relations and practices of power and ask how these relations are present in the more deeply ingrained yet subtle ways in which history and class influence farmers. This approach is arguably more holistic in that it places culture and conceptions of culture, evidenced through stereotypes, as central to family farmers' daily realities.

32. This ideology is one of the tenets of the MST's platform, in fact, and the group has long been known as being against genetically modified organisms. See Stronzake and Wolford (2016) and Motta (2016).

33. Flachs and Richards (2018) discuss the performative element of farming, how farmers must improvise and adapt to dynamic social and natural conditions, values, and aspirations. Flachs also elucidates how knowledge is "cultivated" through relationships, skills, and interpersonal reflections (which Maisa and Carmo reveal throughout this chapter), noting that farmers not only improvise in the field but also perform and practice *mêtis*, subtle shifts they make in relationship to NGOs, associations, neighbors, partners, and even family members, as well as in relationship to the feedback from natural processes (2019, 55).

always moderated by Alba, who places calls, sends emails, and facilitates our face-to-face meetings by driving Carmo, via motorcycle, wherever she needs to go. While Carmo is thoughtful, philosophical, and open, Alba is sarcastic, skeptical, and guarded but a softie at the core. "She's more sensitive than I am," Alba explains with a loud laugh, a wink, and a nod in Carmo's direction. Carmo rolls her eyes. Over time we develop an unspoken agreement that we are in this process of discovery together: Carmo is the sage, Alba the facilitator, and I the ever-curious anthropologist.

Their lives, and their stories, are dynamic. One day, three years after we first met, Carmo hands me a large, professionally packaged basket filled with the latest trends in organic agricultural products from the region—jellies, chocolate bars, liquor. She has assumed the presidency of a cooperative that supplies the cacao for these finished products. She proudly hands me her card and informs me, "I'm on email every day," as she tends to her growing business. "Let's keep in touch."

In subsequent years, Carmo tests out new processing techniques for cacao products. She starts one partnership with a rural association, another with a national agricultural federation, all in the effort to make her production techniques more efficient.[34] Each time we see each other, I think of how we first met on that day, long ago, and of how she had stressed then, and every day since, that "there are no days off in this life."[35]

Her words echo what Maisa often emphasizes as she, and others in the Frei Vantuy community, doggedly run after a better life—an endless pursuit. Maisa has said countless times to me, "I am here on the settlement in the hopes of having a better life. We already have a house. I want my children to live with the peace of being able to say, 'Tomorrow I know I'm going to eat, if I need a doctor, I can see one.' Everything is precarious, and without fighting for more, you will never have anything." From computer skills to organic certification of their crops, residents of Frei Vantuy realize they must be competitive.

---

34. This aligns with Thayer's point on how feminist activism arises in the Global South through processes of interconnection, relations, and "webs of connections." In describing activism and social movements, she notes: "Their politics—goals, strategies, structures—are defined through their engagement with others; there is no essence that exists outside these connections. Social movements do not *have* relationships, they *are* relationships; a set of always shifting interactions with a variety of allies and interlocutors, whether individuals, organizations, discourses or other social structures" (Thayer 2010, 6, author emphasis).

35. See Nogueira et al. (2019) on the need for, and high cost of, labor for cacao production and how this can inhibit family farmers.

They welcome anyone and everyone who can help them. Over time, a regional NGO, led by a strong feminist named Mera, formally establishes its office in one of the thirty-nine homes on the land reform settlement. People hope this will give them more technical training, more access to ways of getting ahead.

And then there are the transformations, almost shape-shifting of sorts, that successful family farmers must master to penetrate the institutions of power—banks, universities, companies, governmental offices—that will open the doors to get ahead in their lives. Maisa and Carmo alike are aware of their advantages and disadvantages. Maisa notes, "I look Italian," which is, in fact, her family's origin. "*Isso ajuda*, this helps," she admits, implicitly referring to the unspoken racial barriers that exist throughout Brazil. Carmo likes to remind me that when we see each other, she is "cleaned up, presentable," as she calls it. "If one day you catch me by surprise on the farm," she says, "you won't recognize me because I will be wearing boots covered in mud and clothes covered in bananas and cacao. You understand?" Carmo tells of how "getting equality is a challenge," meaning getting access to people, to lines of credit, to opportunities. "We are not large producers, without the knife, earning well," she says, recognizing a reality that places her, Maisa, and millions of others like them in a delicate position. As both Maisa and Carmo like to say, they work tirelessly to "elevate the reputation of family farmers." And sometimes for access, you need to look the part. To transcend class and gender constraints, sometimes they need to "hide the knife and boots."[36]

I want to take Maisa's and Carmo's pictures, and the next time I see each of them, I ask for their permission. Both agree, though they have different approaches. "Take it now!" says Carmo as she picks up a large knife to jokingly portray the "stereotypical farmer" while her sister Graça and others cheer her on. Maisa takes a less sarcastic approach. We schedule a time to meet in Ilhéus, and she appears wearing an even nicer outfit than her usual, well-groomed style. I tell her she looks particularly fancy and ask what the occasion is. "Of course I'm dressed up!" she replies. "You're taking my

---

36. Seemingly straightforward situations, such as being a family farmer in Southern Bahia, are often constrained by the ways in which history, class, and culture come together to create stereotypes. But while farmers confront these stereotypes daily, they also actively resist and refashion them. Looking to the formation of particular subject positions, as well as to the ways people resist these subject positions, reveals how power is asserted and negotiated through people's daily experiences. As Goldstein notes about "everyday forms of 'resistance,'" they are "admittedly largely fleeting, but . . . important nonetheless" (2003, 8).

Figure 10 Two representations of farmers, Maria do Carmo (left) and Maisa.

picture, aren't you?" She lifts her cell phone to her ear and instructs me, "OK, you can take the photo now."[37]

Banding Together

On a scorching afternoon, Maisa, Dona Lu, and I sit around Dona Lu's table chatting. Dona Lu and Maisa have always been like mother and daughter. I can't visit the settlement without seeing them both, usually together in Dona Lu's kitchen, where we drink tea and she insists we sample whatever delicious meal is simmering on her stovetop. I have often witnessed the strength each gives the other through listening, advice, and unwavering friendship.

---

37. Maisa's and Carmo's presentations of themselves merge with broader discussions of class as well as gender. When grassroots activists, many of whom are women, gain access to structures of power, how do these processes shape not only class but also gender stereotypes? Maisa, along with many other female farming leaders I engage with, regularly dons high heels to break the gendered barriers of federal agencies like INCRA, the state secretary of education, or the mayor. Such practices invert both class and gender constraints; not just the presence but also the active participation of women in spaces and structures of power formerly closed to them work to position them as subjects in new ways (see Das 2008 on violence, gender, and subjectivity).

My daughter Maya toddles about, going between the house and the garden beyond, playing with pots and pans on the floor, then stopping to venture outside to chase the flock of chicks that have just hatched. There are ample topics to discuss, both serious and light. Dona Lu isn't sure if she should get married to her live-in boyfriend, Antônio. He is almost twenty years younger than her, and while they live together, she is somewhat concerned with what marriage might mean. She is one of the only women on the land reform settlement who owns her plot of land alone as opposed to owning it with a husband. "Does he just want my land?" she wonders out loud to us as we sit at her kitchen table. We grow silent as Antônio walks in and out of the room, seemingly trying to catch our conversation.

Maisa, too, has her own issues to be worked out among friends over the kitchen table. She needs advice on the bureaucratic aspects of leading the land reform settlement and also faces the challenges of balancing her leadership role and her mothering duties. "Don't mind anyone who says you can't do this all," Dona Lu counsels Maisa. Dona Lu is a wise sounding board for all of us: she says exactly what she thinks, doesn't mince words, and offers her uncensored opinion with a loud, scruffy laugh—all while washing dishes, cooking a meal, or cleaning her house.

We chat, waiting for some special visitors who are supposed to come this day. Sure enough, I hear a motorcycle horn out front, below Dona Lu's house, and know they have arrived. In less than a minute, Carmo and Alba appear trudging up the hill, helmets in hand. Carmo lets out a low whistle and says, under her breath so only I can hear, "So *this* is the famous Frei Vantuy!" Alba chuckles and looks around, wide-eyed, taking it in. "Let's get the tour," Alba orders, and together we start off around the property to learn about it all—fruit drying, sheep, education, the NGO, cacao stands on the side of the road. There is a lot going on here. For now.

Suffocating from Cacao

Maisa has disappeared—*sumiu*, as they say in Portuguese. After days of unanswered calls and inquiries to mutual friends, I suddenly get a call from an unknown number. "Meet me in São Jorge hospital in Ilhéus," Maisa says without explanation and hangs up. I instinctively hop in the fuscinha and start the hour-long drive.

Reaching the ancient, poorly maintained public hospital, I walk through the crowded waiting room. While several people look like they are near death, others look like they want to die. I walk up the narrow, dirty hallway that is painted in a tired, shabby green and find the room I am searching for. Maisa rises from the foot of the bed, gives me a big hug, and bursts into tears. She launches into a long explanation of why she is here, gasping for breath as she struggles to communicate her maternal despair.

"Fernando is very sick," she starts, referring to her youngest son, who is eleven years old. "He has these attacks that seem to be asthma, but they aren't really." Her voice trails off, and she looks over at the thin boy lying on the bed. Fernando looks out the window, listless and expressionless. I am not sure if he is listening or daydreaming or trying to coax himself back to sleep. Maisa continues, "It is horrible, just horrible. He stops breathing in an instant, and the only way I can revive him is to do mouth-to-mouth resuscitation immediately." She sighs. "Because of this, I can never leave his side. Never. I can't even really sleep. What would happen if I fall asleep and he has an attack and I'm not around?" she questions rhetorically, looking at me with wide, terrified eyes.

I sigh, welling up with tears in empathy with her helplessness. There is nothing to do for now, other than wait. Maisa has seen doctor after doctor, and the out-of-pocket fees of the hospital visits will set her back months. Plus, the best hospital she can get into in Ilhéus demands an incredible amount of logistical support on her part, for a bed at a hospital is just that, a bed. There are no sheets, food, or even potable water. Her plight with the regional public healthcare system reminds me of a recent experience of ours when I had returned home after a long day to find Max screaming with a bloody cut on his foot. "I stepped on glass in the road," he confessed, and it looked deep. As Jeff was visiting, we had loaded Max into the car and headed to the town hospital. There were a few folks in the waiting room staring blankly at a black-and-white television housed in a metal cage bolted to the wall. Max stopped crying and looked around. A cautious child, he seemed to be assessing the situation. Pointing his eyes at a bloody handprint smeared on the light blue wall, Jeff shot me a look of desperation that indicated his panic—that we were with a maimed child in a subpar hospital deep in the tropics very far from a major city. After an interminable wait, a nurse ushered us into a treatment room where Max perched on a bare metal table while we stood. There was literally nothing else in the room save a small rusty cabinet. While we waited for the doctor to assess the seriousness of the cut, Max brought some levity to this situation. He looked at us earnestly and remarked with complete confidence, "I bet they have actual tools here, you know, but they just lock them up!" In time, Max got sewed up by a perfectly capable doctor and we headed home, well aware of the chasm between those who have privileged access to healthcare and those who do not.

Weeks later, Maisa and I meet again in downtown Ilhéus. Fernando tags behind as her ever-present shadow now. The boy is looking better but still needs to be with her constantly. We buy thick slices of chocolate cake and small coffees, then sit down in the window of the busy café to chat while people rush by. Maisa quietly explains what she has figured out, both relieved and dismayed by her son's diagnosis. "I found out this illness is rooted in cacao." I look at her in disbelief. She continues, "You know it's been raining, and our community doesn't have a good place to keep the cacao beans. The old storage spaces below the bagaças need repair—they aren't dry these days." I nod. "So I've been keeping the cacao harvest from my plot stored in the living room of our house." I nod again, still not understanding the trajectory of her explanation.

Maisa continues. "Well, apparently, after several months of this storage process, mold has grown in our house, worsened by the tropical climate here. Now it has reached really high levels, and Fernando is apparently allergic to mold, so having mold where he lives, where he eats and sleeps, makes his body just succumb to the allergens. When this happens, he has an attack and stops breathing." She sighs, looking at me with dismay.

As we go our separate ways, I replay in my mind all that Maisa has endured. I think of how she came to Frei Vantuy in the first place. She has described this long process to me as "winning the lottery." But actually it was an exercise in perseverance coupled with sheer luck. "There was a guy I knew, Zerinaldo, who was a *vereador*, a city council person, so I asked him about his political promises," she explains. "This is how I started to work with political leadership. This is where I started. I'm that person who absolutely asks for help and also the person who wants to help everyone. He told me about land reform in the region, what people were facing, so I went out and gradually started to learn about the situation here. I saw how much people needed, how people were camped along the side of the road awaiting settlement. And while my life was difficult, their lives were worse. I could still pay rent; they were living in tents without bathrooms."

At the time, Maisa had a relatively good job working in the mayor's office, but the minimal salary she earned left nothing at the end of the month. Her stories are harrowing. She tells of times she couldn't pay for babysitting for her two young children, so she would leave them together in their crib, the three-year-old tending to her younger brother of a year, while she went off to work for eight hours. "I cried as I left each morning, hoping for the best, but what else could I do? At least this way I knew they would still be there at the end of the day." As Maisa relays such stories she looks at me with her big brown eyes, explaining how parenting decisions like this come to be and how her decision in this case seemed the only logical choice.

She continues her story. "One day, when talking with a friend, I heard about a new land reform settlement called Frei Vantuy. Though I felt called to political activism, I didn't have a history with a specific group. I hadn't camped on the side of the road awaiting the government to award me land. I wasn't directly part of an organized struggle, like the MST," she explains. "But I had little to lose. So when one of the thirty-nine homes on this Frei Vantuy was vacated, I knew there would be a vote for a new resident. I also knew I would win the vote—I just knew it."

I've heard this story a few times, and each time I imagine it in my head. I picture Maisa walking from house to house, garnering support. As she explains it, "After working each day, I went to the settlement, and I walked from house to house getting to know people here. I talked with them about their lives. I made friends. I listened to their stories, which went a long way. This was how I worked myself into the community." She concludes, "And when the vote happened, I had invested time and energy in getting to know everyone here." She looks at me earnestly. "I'll never forget them reading the results, opening one paper after another, and announcing my name. It was the happiest day of my life."

As I slip back into the present, I notice Fernando quietly sitting beside us, playing with a small plastic truck as we chat. Maisa returns to the present: "We need to get the cacao out of my home as fast as possible," she says, sighing. "The very thing that sustains us is killing my son. But where am I supposed to put it so we don't lose our harvest?" The following day Maisa figures things out, as she always does. She finds a new space and begins to move the cacao. But while this issue is solved for now, other efforts Maisa has gone through to become a part of this community, to make a life for herself and her family, are beginning to rot like the cacao in her living room.[38]

Eroding

A year later, I return to Frei Vantuy, the model family farming settlement, to find myriad issues tearing folks apart and affecting individuals, families, and the community as a whole. People who have once been allies and friends have divided into two factions—one in support of the current leadership and another against it. Former friends are pitted against each other. These are strong, stubborn, opinionated people, and no one is backing down.

On top of this, some of the settlement's model projects have taken ruinous turns. One project, supported by a state-sponsored program, involves raising sheep. The families of Frei Vantuy pool their labor, build a fence to enclose the sheep, and develop an elaborate plan to care for the animals on a rotating schedule even throughout the night. But suddenly, strangely, one by one the sheep began to die. I meet Maisa in Ilhéus. She is trying to scrape together personal funds to buy medicine for the animals when she gets a phone call. Hanging up quickly, she starts to weep as she learns that the sheep were dying not due to illness but rather because someone inside the settlement is poisoning them. "Who would do this?" she asks me in dismay. Then she utters words I do not want to hear, let alone believe. "They say we know those who are behind this," meaning this is an inside job by fellow community members, possibly even friends.

The chaos does not stop with the sheep but extends to another project in which the community has invested even more time, effort, funds, and hope. One day, as we are hanging out in downtown Ilhéus attending meetings, Maisa answers her phone and looks at me with alarm. "Oh, my God, I need to get back to the settlement right away," she says frantically. "I can't believe it—they locked people in the fruit-drying building. This is completely crazy!" In a matter of minutes, we are in the car, racing back toward Frei Vantuy.

---

38. Maisa and Fernando's story aligns with Lyons's (2020) innovative and creative ethnography on the ways in which soil decomposition, farming practices, and politics converge around the complex context of rural development in Colombia.

As we drive, I remember all that has taken place to establish this fruit-drying project in the first place. I have been following the process for over a year. First, community members had to get a machine installed in a former cacao storage area under one of the bagaças—they worked to secure funding from an international donor, to find a mechanism to safely transfer the money to buy the machine, and then to renovate the space for the processing plant through ardent communal labor, construction, and painting. The day the machine arrived, Maisa and the other leadership proudly gave me a tour of the space. "This is our future," they declared. And, for a time, it was. They worked with the municipalities in the region to secure governmental-community agreements for Frei Vantuy to supply dried fruit products to public schools. Then they established a complex system for collecting, drying, and packaging the fruit according to health code. Hair nets and white clothing were donned by all, with ample sterilization procedures as the fruit passed through the multistage production processes. Perhaps most importantly, there was a system for dividing the labor and the profits among the community members involved. And now, the model project has abruptly stopped with some community members barricading others in the fruit-drying factory.

Maisa arrives with keys to free them from the factory, but the incident leaves indelible scars. It is a marker of how fractured the community has become. If it was difficult to advance when united, it is almost outright impossible to do so when stark divisions exist among those living here.

Bearing Bad News

I collect myself at the table in our second-floor apartment while Guiomar fries chicken next to me, oblivious to the oil that is spraying the walls, as she carries on a loud monologue. The children look bored, tired, and hungry as they draw and wait for our dinner. In the street below us, loud music blares in through our open front door, which also lets in the soft early-evening light. I look down at my phone and see that Rui is calling. This is not a social call; he always has a reason. I escape to the narrow balcony outside our apartment and sit in the hammock in an attempt to find a modicum of quiet amid the chaos, but the early-evening sounds of Itacaré have already commenced. As the sun drops rapidly in the sky, cicadas chirp, frogs croak in the marsh just down the street, and faint music comes from a nearby apartment, punctuated by occasional shouts and the clamor of pots and pans banging as someone cooks. People walk below me on the street, returning from the beach. They inevitably glance up when they hear my voice. Privacy is unheard of here.

"Have you heard the news?" Rui starts cautiously. "What do you mean?" I ask. "There has been a murder at Frei Vantuy. I don't know who it was," he continues. "Oh, my God," I utter, sinking deeper into the hammock. "What happened?" He knows no more. Stunned, shocked, scared, I slowly walk back into the chaos of our home, wondering if it is Maisa who has died. In a few hours I get a text from Rui: "It isn't Maisa. It is another leader on the settlement. The treasurer, I think." I know exactly who he is talking about. I met her the first day I set foot on the settlement and have continued a casual friendship ever since. Her name is Genilse.

Over time, the story of the murder emerges in bits and pieces. It is hard to comprehend. Something doesn't fit. On a Tuesday morning in August 2011, Genilse was at home alone. She was apparently doing laundry, going back and forth from her open kitchen to the backyard beyond to hang her clothes to dry, when late morning, someone entered her home, beat her head repeatedly with a hammer, and killed her. That was it. Her seventeen-year-old son returned home later that day from school to find her. Strangely, though the homes at Frei Vantuy are positioned one after another in neat rows, no one near Genilse's home heard or saw anything. No one had information for the police. There were no suspects, nor was a killer ever found. And there were no clues, save for the hammer, found in a nearby field.

A solemn, almost creepy, picture of Genilse's face haunts the regional papers for days. The community puts up posters calling for "Justice for the Family of Genilse." There is even a rumor that she knew the killer, that he or she was somehow connected to Genilse's life. There are attempts to raise consciousness, to get the police to keep this murder at the forefront of their minds rather than grow tired and move on, as tends to happen in a place where justice systems are overtaxed, where resources and the capacity for a true investigation are limited. Folks from the settlement and beyond, religious leaders, citizens of Ilhéus, even Rui and others I know participate in a *manifestação*, a vigil in Genilse's honor. They march in unison along the road together one Sunday afternoon, calling for peace. These acts, though significant, produce no new leads or apparent interest in the murder.

Throughout this all, I cannot forget how I met Genilse on my very first visit to Frei Vantuy with the professor Salvador. I cannot forget a special trip to her church that we made together. As people prayed in tongues around us and candles blazed in a small rock structure, we formed a bond—she trusted me enough to show me this, and I trusted her to guide me through it. I cannot forget Genilse's wistful smile, her quiet tenacity. I am also haunted by a photograph I have of her: in it, she holds up a hammer, not dissimilar to the weapon that killed her, as she playfully presides at the community meeting we participated in on that first visit to her community long ago.

Staying in Place

Mera, the NGO leader who has moved into the Frei Vantuy community, has aged since I first met her years earlier but still holds her beauty and presence, with neatly braided hair, a gentle voice, a piercing gaze. She has integrated herself well here, and there seems to be mutual benefit: she has a place to live, and, in turn, the community benefits from her leadership skills as she helps people get organized through workshops and such. Above all, she works to promote gender equality and economic opportunities for women.

On a warm night at about nine o'clock, Mera is sitting at her desk, as she does every night at this time, looking out over the field that separates her abode from the highway just beyond. Her house, which also serves as a community office, lies on the edge of the land reform settlement. It has easy access for those coming to meetings and a nice view. Given this, it has quickly become a central hub of activity for some of the settlement's leaders who gather around endless cups of coffee to sort through projects, figure out debt repayments, and try to get ahead.

On this particular night, though, it is eerily quiet. Genilse's death has left people afraid. Fear and mistrust rule here now, and people stay in their homes with their doors locked. Normally, Mera opens the wooden window that looks out above her desk to catch the night breeze. But tonight is different. She works for a bit and then, uncharacteristically, decides to make some tea and lie down. She has a bad headache and she can't concentrate, so she rests on her bed quietly. Suddenly, at eleven p.m., a bullet pierces the exact place she sits working every evening and lodges in the wall behind her desk.

Maisa calls to tell me about this the next day. "The police came. They tried to investigate, but what can they really do? This was someone who knows her, obviously. They even knew her schedule," she admits. People living together in close community over the years come to know each other well. Word travels. Gossip becomes truth; truth is manipulated into gossip. Things take root and fester, and then suddenly there is a bullet intentionally shot late at night. To think that this is the solution for internal community conflict is unsettling. Especially after Genilse.

I meet Maisa secretly in Ilhéus, knowing that Jeff will freak out if I go to the land reform settlement. "What are you going to do?" I ask. "What *can* you do?" She admits, "I'm scared. I'm really scared. I lock my doors. I watch to see if anyone is following me. But what else can I do? I don't have any other land. I own *this* land at Frei Vantuy, and I can't give this up and move to the city. I would have to pay rent. I would have to find a job apart from my plot and leave my role as the president of the association. I have no other options other than to stay here." She pauses, looks out the window at the people hurrying by on the sidewalk wearing city clothes, high heels, and business suits. It all seems trivial, almost ridiculous, against the gravity of our discussion. "I've had threats on *my* life, too," she says. "But what else can I do? I need to stay on the settlement, continue my work, and pray that I'll be OK."

Seeking Justice

For months, nearly a year, a banner calling for justice hangs along the highway where cars speed past, just in front of Mera's house. On a sunny weekend day, the community holds one last demonstration calling for Genilse's murder to be solved. Speakers are set up beside the stage where they convene so their voices are heard. There are songs and speeches urging people not to forget Genilse. They are fully mobilized in support not only of justice but also of respect, of the same investigation and attention the murder of an elite member of the region would demand. At the gathering there are indications of deeper dynamics. One says, "They brutally killed the land reform settler, Genilse. The guilty continue free, inside here." Below it hangs another banner declaring, "Coronelismo, rule by the powerful elite, is infiltrating the process of agrarian reform."[39] And on the stage are about twenty young children singing a beautiful song, hopefully oblivious to the dark reason behind the gathering.

Bahia is a place where crimes can dissipate with the right connections. People know that those with access to those with social and economic power can make problems, even murders, disappear as life goes on. "There has been no justice," people lament to me. They continue to leave the banner up as a reminder that the killer is still at large. We all know there is more than a brutal murder here—there is more that lies beneath the surface. The real reason, beyond losing a friend, is that a model land reform settlement has fallen into a state of chaos. I fear, as does Maisa, that this seems to indicate that family farmers can't be successful, that they can't organize themselves, unite, and overcome the myriad structural barriers to making a living, to building a dignified life—even when, as with the Frei Vantuy community, they have the conditions for everything to go right. "We need to keep Genilse's memory alive," people say. "She is more than just who she was as an individual. She could be any one of us trying to get ahead here."

---

39. *Coronelismo* is "untethered rule by local political bosses" (Mainwaring 1999, 69), synonymous with "clientelism" (Woodard 2005, 100).

Striving for Land and Livelihood

How do subject positions take root and persist to shape social movements and the people involved with them, and how do people exert agency and strategically manipulate beyond particular subject positions as they strive to better their lives?

Although the cacao era has passed, Southern Bahia still bears the imprint of deep divisions of class and culture. Family farmers throughout the region must navigate complex politics of subjectivity: their livelihood efforts can be constrained when others associate land reform and farming with entrenched stereotypes about poverty, laziness, violence, and environmentally destructive agricultural practices. At the same time, farming looks different today as former wage laborers and migrants from cities strive to take advantage of new opportunities through land reform initiatives and growing markets for organic agricultural production. This chapter explores how subject positions are constructed and perpetuated, as well as resisted and refashioned. Attention to subjectivities can help us better understand both the limitations and the power of social movements. Here, amid these struggles, farmers' stories of striving reveal the complex dynamics of power they must navigate, as well as their hopes, dreams, and very real attempts to create better lives for themselves and their children.

Finding Present-Day Voice amid Historical Silence

"The mayor was supposed to come here. I got everyone together, we sat waiting all day. But he never came."

—Jitilene, Itacaré, Bahia

Fleeting Presence

I am drawn to the Rio de Contas, a mighty river that begins in Bahia's westernmost part, descends for nearly 400 miles, and finally passes by the Porto de Trás neighborhood of Itacaré to empty into the Atlantic Ocean. One of the best views of the river is at the top of a short, steep hill that leads up from the banks of the river to the Porto de Trás Cultural Center, a relatively new structure financed by the Swedish rock group ABBA. From the cultural center, if you look very carefully through the thick Atlantic Forest vegetation, you can see the river flowing far below. Seaweed from the ocean floats upriver in high tide, downriver in low. A canoe might slowly drift by with two or three figures quietly casting fishing nets out over the water. A small motorboat laden with people can sometimes be seen coming to or going from town.

Sometimes, from high on this hill, I am graced with a glimpse of one of the elusive river dolphins that reside here as a smooth gray mass rises out of the dark water momentarily and then slips back in. If I am not looking at the exact spot where the creature serendipitously surfaces, I will miss it. There might be other evidence of this presence—a few bubbles floating on the surface of the water or a barely audible spouting noise. But most of the time

I must be content just knowing that the dolphins are here. On a sweltering summer day in Bahia, I walk down from this viewpoint, toward the center of the town, thinking about how fleeting indications of reality sometimes have to be "enough."

Meeting Dona Otília and Jitilene

My companion on this blazing hot afternoon, Dona Otília, chats with me in the town square. Papai Noel, or Santa Claus, as he is aptly named by tourists due to his long white beard, picks at his post-lunch food remains, carefully cleaning his teeth with a toothpick. A few mangy dogs lounge in the sun as flies buzz about their greasy coats. I contemplate purchasing a fresh coconut to stave off my thirst. Raul, the owner of a small inn at the mouth of the Rio de Contas, walks by. Dona Otília, ever aware of all taking place around her, stops our conversation about quilombos in the region and calls out to the man. "Hey, did you know that Itacaré used to be one of the largest quilombos in Brazil?" she asks, referring to communities of descendants of formerly enslaved people.[1] "Eishe!" he exclaims

1. Quilombo is the term used to describe a community comprised of Afro-Brazilian descendants of formerly enslaved people who fled slavery, and quilombolas are the people who live in quilombo communities. Quilombos are also known as *mocambos*, maroon communities, negro lands, palenques, and most officially, as *comunidades remanescentes de quilombos*, or "surviving quilombo communities." João José Reis notes, of quilombos in the region, "Taking advantage of a region that had few people and was little guarded, slaves formed mocambos [quilombos] since at least the seventeenth century in Camamu, Cairu and Ilhéus" (1996, 339).

Quilombos often did not exist isolated from the rest of Brazilian society, although they were perceived to be separate in the popular imagination. In many cases, they were, and still are, deeply integrated with other communities (Castro 2006; Leite 2012; Linhares 2004; Luna and Klein 2004; Mahony 2001; Schwartz 1996). Quilombolas in Southern Bahia often forged cross-community relationships, relying on white landowning farmers who supplied plots of land in exchange for their labor (Reis 1996). Mahony's (2001) historical research reveals how Afro-Brazilians in Southern Bahia at the turn of the eighteenth century both had long-standing ties to quilombo communities and were also integrated into the broader agricultural and commercial scene by serving as farmers and merchants.

However, even with these historical and present-day relations among quilombolas and towns throughout Southern Bahia, the dominant perception of quilombos as far-off, isolated communities still exists among Brazilians and can shape conceptions of what constitutes an

with a low whistle. "I had no idea at all!" "And your business lies at the gateway to these communities!" Dona Otília chides him, referring to the small pousada, or inn, that Raul runs for tourists. "You gotta learn your history, man!"

Dona Otília is an octogenarian badass. She is slow to smile, much less laugh. Given the quick pace of her mind, the rapidity of her speech, and her matter-of-fact manner of talking, she is often outright impossible to understand. This is maddening, for she is indubitably a person worth listening to. Dona Otília is brash and abrupt with people she doesn't know, those she couldn't care less about, or those who need to be put in their place. Most often, this is everyone around her. She intimidates folks who are hard to intimidate, like Guiomar, who questions me when I give her something to deliver to Otília's home, which lies at the entrance to Itacaré. "You're gonna make me go up there again? That's a mighty powerful woman, you know!" Guiomar says. But at the root, we all know that her bark is worse than her bite; her brazenness is worth its price. Dona Otília is one of the most knowledgeable leaders in this part of Southern Bahia— and has been for well over half a century. She has distinct ideas of what is right and what is wrong. And she is audacious in a place that can benefit from more audacity.

Dona Otília has a distinctive way of moving through the world; she is a font of perpetual energy. Those who want to truly accompany her, physically as well as mentally, must "walk by her side," as they say here. In her home, Otília shuffles about, acting as the matriarch to her massive extended family of more than fifty. She barks out orders. "Get that mop of hair brushed," she instructs her young granddaughter. "Fry me an egg, you lazy one," she tells her son. She sweeps, then prunes her plants on the ledge of her humble home, then turns to the person visiting her, never missing a beat. "OK, let's get on with our chat now, shall we?".

Otília moves about Southern Bahia, meeting with everyone from mayors and university leaders to farmers and elders. And she is always, always, on the side of the underdog. "My life, since I was twelve years old, was about fighting for someone. I left my parents in the countryside of Bahia and

"authentic" quilombo. Leite (2015) argues that the meaning of quilombos can shift depending on the social, political, or historical context and, in a dynamic, multicultural, multiethnic Brazil that is a vibrant hub of Afro-Brazilian culture, quilombos can still exist at the margins of the country's broader Afro-Brazilian identity (see Ferreira 2006; Véran 2002).

stowed away on a boat to the capital of Bahia, Salvador," she once tells me, as if this explains both her independence as well as why she has been fighting ever since.

Throughout decades of being a social activist, Otília has served as a *vereador*, a city councilperson, an advocate for land rights with agrarian reform settlements, and a champion of causes ranging from gender equality to educational improvement. When Otília decides to do something, it almost always happens. Well over two decades ago, she decided to take up the cause of quilombo communities and quilombola people. Quickly, information about and relations with quilombos and quilombolas in Southern Bahia began flowing to, through, and from her. Though Otília doesn't live on a quilombo herself, her mixed bloodline gives her plenty of credibility to represent historically marginalized populations.[2] "I'm a mixture of Afro-Brazilian, *caboclo*, and indigenous Tupí-Guaraní," she proudly states. "I'm a complete mutt, and proud of it."[3]

In contrast to Dona Otília, another leader in the region I came to know, a young woman named Jitilene, initially seems the polar opposite. Jitilene, who is merely in her twenties, is shy and seems almost reluctant to speak at times. When she does, she looks down or to the side and speaks quietly with an air that can almost be mistaken for indifference. Perhaps this is not a mistake but rather is intended.

Jitilene lives a steady twenty-minute canoe paddle up the Rio de Contas on a quilombo called Santa Amaro. They know each other well, for Otília is friends with Jitilene's mother. As Otília describes Jitilene's community, "Of the quilombos in this region, Santa Amaro is the poorest," which is particularly sobering, as Otília herself lives in very humble conditions. And of the communities throughout Southern Bahia—fishing villages, farming settlements, towns, and cities, among others—quilombos are arguably the most

---

2. Cossimiro, president of one of the local quilombo associations, explains, "Otília has a lot of history here. I think there is no one who has better knowledge of [quilombos] here in Itacaré." Otília wrote a book on the history of Itacaré that Floresta Viva published in 2009; she has also written several volumes of historical and political poetry, and frequently recites her poems passionately in public meetings.

3. A *caboclo* is a peasant of mixed Brazilian Amerindian and European ancestry who lives in the countryside. This term originated in the Amazon region (see Nugent 1997). Tupí-Guaraní are Indigenous peoples who were living throughout Latin America at the time of colonialization and who still reside in Bahia today (Almeida and Neves 2015).

impoverished. There are seven quilombos in this region, yet many of these communities are little more than a cluster of barely standing, ramshackle houses. The people living in quilombo communities often exist at the margins of survival. Food can be scarce, basic infrastructure is frequently absent, as are other important social systems, such as transportation to schools and jobs. Quilombolas, the people living here, often lament, "No one comes to our communities. They are completely forgotten, completely *jogada*, cast off, without water, without jobs."[4]

Quilombos are often perceived as fleeting remnants of Brazil's African past. People seem to expect to arrive at these communities and be hit with visible and tangible evidence of a bygone era, with formerly enslaved people cooking over a smoldering fire and perhaps singing and dancing around a samba circle for diversion. These antiquated and romanticized notions, however, are almost conspicuously absent in present-day quilombos. These communities are far from sites of fixed memories of an idealized past. Instead, many quilombos are characterized by processes of forgetting within the context of contemporary Brazil, existing as liminal spaces where people live their lives "in between." One day Jitilene explains her community patiently to me, "Out of thirty-three quilombola families, only thirteen actually live on the quilombo." This in-between state can be not only symbolic but also can have material and even larger structural consequences. Jitilene continues, "So I really want to get a boat project going with the mayor because right now kids can only study in the morning since there is only one boat to town. The mayor doesn't pay for children to study in the afternoon, so they wake up very early and attend school, and then head home around lunchtime, or they stay in town with relatives."[5]

At twenty-one years old, Jitilene has recently been elected the president of the Santa Amaro quilombo. She explains why this happened, despite her young age: "I'm the oldest one who can read and write, so I need to do this."[6]

---

4. As Linhares notes, "Throughout official history, more precisely after the abolition of slavery, these social groups had been forgotten" (2004, 828).

5. Many quilombolas in the region lead lives of constant mobility between the towns that provide employment opportunities and their *roças*—small plots of land for subsistence farming in quilombos. People often leave their communities each day or spend an entire week away, sleeping in nearby towns.

6. Becoming a recognized quilombo demands that people form an association to access government and civil society funding. Jitilene has the energy and tenacity to pursue the myriad

Figure 11 Jitilene paddles up the river to visit her quilombo community, Santa Amaro.

In fact, Jitilene needs to do a lot. She has a seven-year-old daughter named Eduarda to care for. She has to finish high school. She has to navigate between her life on the quilombo up the river and her time in town in a home her mother, Dona Julia, owns in Porto de Trás, one of the oldest neighborhoods of Itacaré, which is known as an "urban quilombo."[7] On Sundays, Jitilene also needs to help her family sell homemade liquors at the town fair—all while maintaining a constant leadership presence. "Nearly every day, I paddle to

bureaucratic processes necessary for her community to become a formal association, which often involves chasing after people for signatures, taking trips to the urban areas in Bahia, and doggedly fighting for rights and services for the quilombo. At the same time, there are limits to participation in these processes, which demand technical knowledge, specific language and discourse, and even the time and ability to pursue these strategies.

7. Brazil's first "urban quilombo" was officially designated in the city of Porto Alegre in the southern Brazilian state of Rio Grande do Sul in late 2005; it is located among mansions and high-rise condominiums (Leite 2015, 1236).

the quilombo to check in with people," she tells me, reaffirming her commitment to her post as president. But at the core, Jitilene is not that different from Otília. She wryly states, "I'd rather fight for the quilombo every day than wash dishes for someone else!"

Fighting for Memory and Culture

"Fighting for the quilombo" regularly takes place on the streets of Itacaré—every Monday, Wednesday, and Friday, to be exact, in various spots, both public and not. One of the best places to witness this is arguably O Xaréu, the viewpoint, which lies at the top of a small hill looking over a meeting of the waters, where the Rio de Contas enters the Atlantic Ocean. Just a few kilometers down the river from where quilombos were first established in Southern Bahia, people gather nightly at O Xaréu to watch the sunset as well as the capoeira. This literal and figurative fight intertwines song, dance, drums, and, most of all, movement as mostly Afro-Brazilian bodies gyrate in concert as they anticipate their partner's moves and, in turn, react to fool, overcome, or flee the person with whom they are fighting—or "playing capoeira."

Overseeing the capoeira at O Xaréu and other locales around town, from bars to open areas in the town plazas, is Master Jamaica, often simply known as "the Master." His manner is rough, like Otília's. He scorns those under his watch, pointing out their weaknesses and publicly poking fun of them. They take it all willingly, silently; there is no other choice.[8] He is the Master. Jamaica presides over a group of mostly men, whom no one calls by their given names. Instead, he has baptized them with new names that fit some aspect of their personality, their physical body, or their way of moving in the world. Guiomar's son Alberto, who left Brazil a few years ago, has become Neto, or grandson, after his father's and grandfather's lineage. Andre has become Manso, for his calm or soft nature. Robério has become Nino, meaning little, an ironic nickname, for he is well over six feet. While these capoeiristas were born and grew up in the region, they have now transferred their skills to Europe. Neto is in Sweden, Manso in the Czech Republic, Nino

8. See Kurtz (2020) on women's gendered attacks in capoeira and "guerreira" tactics in response.

in Norway. They come back once a year to bridge their lives over there with their memories, their training, and their home, which will always be here.

And there are others. Júlio has become Comprido, meaning simply "Long," to describe his lengthy dreadlocks. Tião is now Caboclo, a person of mixed Indigenous and European ancestry who most often lives in rural areas, to indicate his roots. Both of them have either gone away and come back or found a way to live between two worlds. Despite—as well as because of—the ways of the Master, they have found a strategy for making life out there, la fora. And this life has been intricately shaped by capoeira, the fight disguised as a dance to unsuspectingly rise up against slavery, the fight that bridges past and present, the fight that, in many cases, gave rise to the possibility of quilombo communities rising up in the first place.

Constructing a Concept

The first time I visit a quilombo community, I tag along with members of Floresta Viva to observe a workshop they are holding. The NGO staff has spent days working with the quilombolas to assess the community's needs and decide how they can best develop a project in collaboration with the people living here. Now they are presenting their results in a workshop. After the long day, everyone is spent. People loll about on the front porch of a house, waiting for the canoe to transport us back down the river to the closest town. To pass the time, Sayonara, a young woman who works for Floresta Viva and serves as the primary outreach person with quilombos, asks people to sing a samba.[9] The quilombolas either cast their eyes downward or are quiet. Some dart furtive glances or barely distinguishable smiles at each other. All are silent; no one willingly offers up a performative song upon her command. "Come on," Sayonara urges, "sing us a samba." Breaking the awkward silence, one brave soul offers an explanation meant to placate her: "Agh, we've forgotten the words." People shift their feet and look down again. She will have none of their excuses and again cajoles, "Sing one for me!" She starts to clap her hands and tap her feet rhythmically.

---

9. Samba is commonly heard on the streets of Bahia. What they are referring to here is a particular style of samba, called a *samba de roda* (dance circle with music and poetry), that is typical of this region.

Finally, one man obliges and starts singing in a low, tentative voice. Others slowly and quietly start to clap along, joining in. Someone begins to beat a rhythm using a board and bucket lying on the porch. Soon, a somewhat weak rendition of a samba, but a samba nonetheless, gradually, tentatively begins to rise up, accompanied by the sounds of cicadas and frogs as night falls in the forest.[10] When the motorboat starts up the small inlet to take us back to Itacaré, people abruptly stop singing. The samba instantly becomes a memory. Later, when I recount the call for an impromptu samba at the Santa Amaro quilombo to Dona Otília, she scoffs loudly, "Samba, samba, samba—people think quilombos are no more than this!" What is behind the reluctance I witnessed in conjuring a samba? When official histories are forgotten, or unknown, what can come from acts of remembering? I think about fleeting evidence of presence, like the river dolphins.

Later, I pose a simple question to Sayonara, who had insisted on the samba in the first place: "What makes a quilombo a quilombo?" I know that quilombos are communities of descendants of formerly enslaved people, people who fled one of the ugliest institutions known to human history. I also know that in 1888 Brazil formally abolished slavery, the last country in the Western Hemisphere to do so. But how do these communities of former escaped enslaved people exist in the world today? How do the people living here claim a connection to the past as well as an identity in the present? Sayonara launches into a response, methodically listing some of the characteristics of quilombos. "Well, for one, you have evidence of where *fornos de lenha*, wood fires, used to be," she begins.[11] I think of

---

10. While Sayonara was trying to animate the group by calling upon a shared history, this calls to mind anthropologist John Collins's (2015) eloquent portrayal of the myriad ways in which "performing culture" is carried out by residents of Salvador's Pelorinho district, a World Heritage site, as people themselves are appropriated as a form of cultural patrimony. It also aligns with Ramos's (1994) discussion of what constitutes identity and the concept of "hyperreal Indians" as those being the most authentic and worthy of defense and support.

Anthropologist Richard Price claims that present-day quilombos "share a different heritage of 'quieter resistance' than their heroically celebrated revolutionary ancestors" (1998, 246). Was this "forgetting" of the samba resistance, per se, or merely forgetting? If the former, in working to combat this forgetting, how are quilombo leaders and their allies not only shaped by the politics of the past but also agents and subjects that can influence the politics of the present (Trouillot 1995, 23)?

11. Farfán-Santos (2015) writes about how the authenticity of quilombo communities depends on quilombolas' abilities to perform and recount their history "correctly," meaning it

how wood-fire cooking is being revived across Brazil: my wealthy Euro-Brazilian doctor friend who lives on a farm outside of town had proudly showed us how his forno de lenha "makes a good pizza" when we spent last Sunday at his home. There has to be more. She continues, "You also have archaeological remnants like pots, you have stories of histories of slavery passed down from elders to the people living in a place today, you have a tradition of *samba de roda*, samba in the round, perhaps a *Candomblé ter-reiro*, a sacred place where the Candomblé religion is celebrated." She trails off and looks out at the river before us as if imagining the vibrant communities of quilombos established here long ago. I take a sidelong glance at her, not wanting to interrupt her reverie. Perhaps, in fact, this is why she attempted to conjure up a samba the other day. And my mind drifts off as I imagine terreiros up the river, for the only terreiro I know, which I had recently visited, lies in the center of town.

Jeff was in town. It was one of his "weeks on" with the family, which happened every five weeks like clockwork. We both needed this time. Connection between two people can be lost when you live parallel lives on different continents. It can be particularly difficult when one life is propelling forward in new and exciting and challenging and unexpected ways while the other is tethered to the quotidian, the familiar, the regular. I was the former; he was the latter.

But this night I had plans that nothing, not even a visit from Jeff, was going to interrupt. I rose at three in the morning, nursed Maya, and strapped her to my chest in a front pack. I didn't want her waking up and crying, which would arouse all in our small house. Maya was my constant charge, fieldworker or not, for mothering is not easily turned on and off, and I had the breasts. In the dark of night, she and I set out through the streets of Itacaré, walking slowly and quietly. Off in the distance, I heard the boom of a bass drum from one of the late-night *forró* parties that had yet to shut down.[12] I paid attention to the night noises, for they calmed me. Frogs, crickets, other unidentifiable creatures. I could not be afraid. I knew if I

---

has been integrated into the Brazilian national imaginary.

12. *Forró* is a dance that originated in the Brazilian Northeast and has spread throughout the country and even to other countries and continents. The origins of forró are disputed. One account suggests that it comes from the English expression "for all," while others say it derives from an African expression *forrobodó*, which means "popular dance" (Quadros and Volp 2005, 117) or "dance of the common people" (Crook 2005, 267).

opened myself up to fear in any way, at any time, while living alone in Brazil with three children, it would be over. I passed the pousadas where tourists rested up for their eco-excursions the following day. I passed one of the few undeveloped forest tracts in the town, where thick vines hung from the enormous trees, their moist leaves glimmering in the moonlight. I rounded the corner and peered ahead, looking for a sign of what I was searching for. Gradually, my eyes found it, a sliver of light emanating from a crack in the door of a small blue house on the corner.

Slowly, I opened the door and entered gingerly. The walls inside were a particular blue, a color that reminded me of the robes of the Virgin Mary, which I knew well after sixteen years of Catholic school. Jitilene's mother, Dona Julia, was at the front of the small rectangular room. She sat on a rough stool in the front corner. A few women gathered about her, helping her in different ways. All were dressed in white, like the Baianas I had seen long ago at the port in Salvador with Mario and Paula. To her left was an altar with candles, images of saints, fresh flowers. It felt familiar. Dona Julia looked up at me with a dreamy expression and nodded quietly, then returned her attention to the women—and it was mainly women, save a man or two—who were getting ready for the ceremony.

A filmmaker from Southern Brazil with long, blondish dreadlocks noticed me as I entered with Maya. "You came, after all, eh?" she noted with surprise. I nodded and wished I could blend in more somehow, but it was impossible, being the only other non-Afro-Brazilian in the small room and with a child on my front. Thankfully, no one was concerned with me. They had work to do. The filmmaker milled about in the back, setting up cameras. As I took a seat on one of the wooden benches, I wondered if she had permission to film what was going to take place and what she planned to do with it.

In addition to Dona Julia, I recognized the woman who was apparently in charge of the ceremony. I passed her house, adjacent to the terreiro, bringing the kids to school each day. On the exterior wall, she had written "cold beer" for passersby who would decide that her home, rather than the gas station just beyond, was where they needed to buy beer. And I often found her beating rugs, hanging clothes to dry in a narrow line that bordered the dusty street, or simply sitting by her front door, as people often do in Bahia, watching the world pass by. I remained quiet. Soon, one of the men began to beat a drum, and the ceremony began.

What ensued was simultaneously familiar and exotic. While I was familiar with religious singing, drumming, prayers, and offerings, the chanting and being visited, or overtaken, by spirits was somewhat novel. On this night in the terreiro, the holy space where we gathered, when Dona Julia and others transformed as spirits borrowed their bodies and voices, I watched with curiosity and reverence. I prayed in my head in English to invoke the familiar in a situation that was not quite so.[13]

After a few hours, we got up and slowly progressed en masse to the sea. Roosters crowed, and the town gradually started to wake up from the night. We heard a few pots banging as people put on morning coffee to brew. Once in a while a wooden window would swing open and the sleepy-eyed inhabitant would look out to greet the day. Maya continued to slumber on my chest, lulled by the gentle beat, the singing, the walking. We rounded another corner and headed toward a small cove along the coast. A man passed, carefully carrying his caged pet bird for the morning walk. Another woman stopped sweeping her porch and paused to watch our procession. In time, we reached the coast, where people continued to drum, chant, or in my case just reverently listen and sway to the beat of the music coupled with the rhythm of the ocean.

As the sun rose beyond the lighthouse where the river meets the sea, we leaned down, offering flowers and fruits to Yemanjá or Iemanjá, the female goddess of the sea. With all my years of Catholic education, I understood ritual. Though the Candomblé religion was new to me, I was eager to gain a glimpse of what had meaning to people in this place. Cradling Maya's head carefully to keep her quiet for a few more minutes, I reached down and placed a red flower carefully in the waves, watched as it was carried out to

---

13. Nearly two decades of formal Catholic education equipped me well for ritual and ceremony, and transformation by spirits was not entirely new to me. A decade earlier, when a friend of mine was called to channel spirits, she screamed and writhed on the ground. Before my eyes, I watched the person I knew turn into others. First, she became a prostitute. "Come here, boy," she sneered at one of our friends in the room, shaking her hips and motioning to her genitals. Then she was an old lady, with a weak, shaky, unintelligible voice. I had watched in quiet disbelief while a mutual friend calmly explained to me, "She has a history of this in her family. It's what they do." As the intensity of her channeling grew, we moved outside to give her privacy. We stood in a circle, held hands, and prayed. My Brazilian friends prayed aloud in Portuguese while I silently said Hail Mary after Hail Mary in English, all of us collectively conjuring whatever spirit or god spoke to us.

sea, and said a little prayer to the God I knew, who most certainly was connected to the God those around me praised.

As the sun rose, the music gradually died down and stopped. The night
was over, the ceremony done. It was early in my time in Itacaré. There was
much I did not know back then. I did not know that Luizinho, the lead drum-
mer, would help me fend off potential evil characters we would encounter
along the regional road. I did not know that Dona Nena, the priestess of the
terreiro, would fall ill and die. I did not know that in a few years the terreiro
would disband, that the beer sign on the home next door would be painted
over, that the peaceful energy there would dissipate. I slowly walked back
through the streets, passing from cobblestone to dirt as I entered our neighborhood. Jeff and the other children were still sleeping. I placed Maya back
down in her crib and quietly slipped back into bed.

With the call of a bird from the riverbank, my attention returns to Sayonara's explanation of quilombos in the region. Were there terreiros nearby?
Did people practice their religion here, up the river? Or, rather, did we need
to be content knowing that they once existed and that descendants of escaped
enslaved people may—or may not—be carrying on their religious traditions
here to this day? I look across the river to the land that has been cleared for

Figure 12 Left to right: Perpétua Maria Conceição with Clare Eduarda, Dona Julia,
Jitilene, and Welber Silva dos Santos, celebrating quilombola culture in ceremonial
Candomblé dress.

planting and the dense forest beyond and think about Oitizeiro. This name has been surfacing ephemerally, again like the dolphins, at the mention of quilombos in the region. But when I have asked people to take me to Oitizeiro, my request has been met by cocked heads and unexplained silences. After months of searching, I have discovered that Oitizeiro, one of the region's most famous quilombos in the past, no longer exists in the present.

"So why do people keep talking about Oitizeiro?" I ask Sayonara, trying to understand why a community that has ceased to exist is still such a contemporary reference. She patiently explains, "Oitizeiro is related to the loss of land, to the act of losing land. The problem is some quilombolas don't know what a quilombo is. Information comes out on the radio, and also through word of mouth, that territories that were once quilombos are now being taken over by people who are not quilombolas. But since people don't know what quilombos are and that they themselves are quilombolas, they end up really losing!" She sighs. "They also don't have documents. So it is a process to construct that they are quilombolas. And sometimes this is an external concept."[14]

I wonder: How do people self-identify as quilombolas, particularly if the notion of a quilombo community of fugitive formerly enslaved people is at times an external concept that has been unearthed from the past, appropriated, and passed along by the government, civil society actors, transnational

---

14. In 1988, 100 years after the abolition of slavery in Brazil and after two and a half decades of military dictatorship (1964–1985), a new constitution was drafted in Brazil. In addition to two constitutional articles that officially recognize the contribution of *grupos negros* (Afro-Brazilian groups) to Brazil's cultural heritage, the constitution also contains a one-sentence provision that provides a foundation for present-day quilombo identity and rights. Article 68 declares that for "the remaining quilombo communities who are occupying their lands and recognized as the full owners, the State would issue them the respective titles."

This constitutional inclusion established a structural mechanism through which quilombo communities could finally emerge from centuries of what was initially intentional obscurity (as fugitive enslaved people) and later figurative obscurity within the broader social and political context of Brazil. With this new constitutional provision, the Palmares Cultural Foundation was established, also in 1988, by the federal Ministry of Culture to identify and designate quilombo communities in collaboration with INCRA. Brazilian historian Hebe Maria Mattos connects this constitutional inclusion to a long process of reconstructing the role of Afro-Brazilians, brought about, in part, by growing Afro-Brazilian activism in the 1980s that made the memory of slavery public. She notes, "The approval of the article about the land title rights of the 'quilombo communities' thus capped an entire process of a historical revision and political mobilization, which linked the affirmation of a black identity in Brazil with the dissemination of a memory of the struggle of slaves against slavery" (2005, 3).

agencies, and foundations? What happens when quilombo identities are unknown, unclear—perhaps even undesirable—to the people who can rightly claim this identity? Can an external concept be strategically appropriated?[15] I also wonder why, somewhat ironically, the historical memory of the quilombo that is now long gone, Oitizeiro, seems to attract more scholarly attention than the present-day quilombos in the region, the places where people still make their lives.[16] And I wonder why particular histories and identities take hold, while others remain hidden or unclaimed.

I know that Rui's NGO, Floresta Viva, recently conducted a study of quilombos in this part of Southern Bahia. But while stories of slavery exist in each community, only two quilombos openly claim they were "formed in the context of slavery," and out of these two, only a few people here seem willing to speak of this. It seems that tracing a past of fugitives from slavery is a complex task, not only archaeologically and historically, but that this process also conjures something that is either unknown or perhaps intentionally hidden. Perhaps these are painful memories, passed down through generations, of past oppression. Perhaps these are reminders of structural inequality that still pervades in the present. I don't know, but I know some people around here do. But they don't seem to want to talk about it.

Silencing Past and Present

Dona Otília and I lumber along in the fuscinha, picking up hitchhikers on our way into Itacaré as I always do. It is a challenge to fit four more people into a small car, but we do so anyway. Plus, Dona Otília knows them, so we cannot pass without stopping. As we drive, she continues giving her history lesson to me, oblivious to the fact that our new passengers might not care. She talks about how she insisted that quilombos come to be recognized as such, as formal communities of fugitive formerly enslaved people. "People

---

15. In her work with quilombo and Indigenous communities in the state of Sergipe to the north of Bahia, Jan Hoffman French asserts that quilombo communities are aware of the ways in which identity can be strategic for making particular claims of land, rights, and belonging (2006, 2009).

16. Oitizeiro was one of the oldest quilombos in Southern Bahia, dating back to the 1600s, and is often touted as the symbol of quilombos' and quilombolas' place in the history of the region and, more broadly, Brazil.

from INCRA and people from Palmares, the federal Palmares Cultural Foundation, said Itacaré didn't have quilombos.[17] But I went to meetings with the leaders of the mayor's office and I said, 'Yes! I'll show you to the contrary.' I brought the map of this region and said, 'Santo Amaro, what is this? Oitizeiro, what is this?'" she exclaims with vehemence, listing off the region's quilombos one by one.

We arrive at the entrance to Itacaré, and she waves to the west where we see miles and miles of dense forest. Far below the sea of green, way off in the distance, I see the Rio de Contas winding through the forest. Otília looks out and matter-of-factly states, "Quilombolas and Indians, quilombolas and Indians, all the way from here to Taboquinhas, a city thirty kilometers up the river." She continues, clearly on a roll now, "Once, we found these people hiding in the forest. 'Come out here!' I yelled to them. They didn't come. 'Come out here!' I yelled again. Finally, they came. I looked at them and asked, 'Why are you hiding?' They didn't answer." I smile to myself, imagining what an intimidating figure Otília must have been with her rough but good-intentioned manner. She continues her story. "So I say to them, 'You were enslaved, weren't you? Tell the truth, people!' And slowly they nodded yes." She looks at me and exclaims, "I told you so, this reality exists here."

Whether the people Dona Otília encountered were suffering from some type of clandestine, present-day slavery or were ancestors of formerly enslaved people, of quilombolas, is unknown. Whichever it was, Dona Otília is soundly convinced that the people she met in the woods that day had either a present identity of slavery or a strong connection to the memory of their ancestors' oppression and that either possibility was so viscerally ingrained, it caused them to hide from her.

This guardedness is found in other venues too. At monthly planning meetings for the APAs in Southern Bahia, quilombolas' presence can be illusory. More often than not, their participation may be marked by a signature on the attendance roster rather than by vocal and engaged discussions and debates on environmental and social issues of the region. Often, after a morning of meeting, when the group of community representatives disbands, I realize the quilombola representatives haven't uttered a word.[18] Such "spaces of

17. Brazil's largest and most well-known quilombo, which no longer exists, was Palmares (see Price 1996; Schwartz 1996).

18. Given Brazil's history of slavery, as well as the racism that persists in the present, how can we look to "spaces of silence" that are present within the structures, relations, and

silence" subtly hint at how history, memory, identity, and power converge in people's lived experiences, as well as how these forces continue to rise up, to be perpetuated, and perhaps, at times, even to be resisted.[19] Sometimes,

---

identities of quilombolas (Scanlan Lyons 2011)? There are myriad examples of silencing, such as in the legal system through bureaucratic difficulties of obtaining designation and land title. Escallón writes of quilombolas who are denied land rights and rendered invisible. She notes, in speaking of a land conflict around a park designation on quilombo land, "the silence regarding the inhabitants . . . is full of meaning" (2019, 374). Michel-Rolph Trouillot claims silencing can take place through erasure, the failure to consider particular realities, as well as through what he calls "banalization," the stripping of context and power (1995, 96). Forms of silencing in general, and erasure and banalization more specifically, are useful ways for considering the silences that influence quilombo communities and quilombola people. To use Trouillot's terms, for example, quilombolas are expected to endure both erasure as well as memory of the painful elements of both their past and present.

　　Furthermore, to understand silences, quilombo communities and quilombola people must be considered as deeply embedded within the historical as well as contemporary politics of race in Brazil. Gilberto Freyre claims that Brazil's colonial experience indelibly constructed its race relations and created a "myth of racial democracy." Although Freyre was writing in the mid-twentieth century, his discussion of how the colonial condition created both racialized and gendered mechanisms for domination is an important context within which to consider quilombolas' existence. Furthermore, while Freyre first raised the issue of race in Brazil, in doing so he also opened a space for connecting slavery, race, and trauma. In a lecture he gave on the subject, he noted: "As a whole, Brazilians have what psychologists call a traumatic past. Slavery was their great trauma" ([1946] 2013, 122). Freyre's point on the difficulty of connecting the realities of people's present to the painful memories of their past provides a foundation for understanding the paradox facing quilombolas today—they must connect themselves to a particular past to assert themselves as a social movement in the present. In other words, to adopt a cohesive and authentic quilombo identity and the rights, services, and opportunities associated with this identity, quilombolas must conjure a collective memory of their enslavement. And to live in the present, they often must balance memories and remembering as well as erasure and forgetting amid the present-day realities of racial oppression in Brazil. Scholars have characterized racial politics in Brazil as both historical and present-day gaps and erasures (see Chalhoub 2006; Collins 2015; Kraay 1998, 2016). Numerous studies since Freyre have also dispelled the myth of racial democracy by unmasking the real-life situations that belie true democracy across color lines (see Collins 2008, 2015; Goldstein 1999, 2003; Htun 2004; Lovell 1994; and Mattos 2005, among many others).

　　19. Situations that I encountered in Southern Bahia provoked me to look more deeply into the lived experiences of quilombo communities and quilombola people as an emerging social movement that is very much couched within the complexities of history, race, class, and power in Brazil. Studies outside the context of Brazil, which explore how "histories and silences are produced and unproduced" (McGranahan 2005, 570), made me curious about this concept among quilombo communities. Drawing from this idea of the production of silences, analyzing

too, there is no hiding the connections to a painful past, for they bring it into the present whether naïvely or quite intentionally. And sometimes, laughing and poking fun at a painful situation from the past is one of the best defenses in the present.[20]

Like the other day. Guiomar pointed at her boyfriend, Manoel's, legs as he carried material she had just bought for her never-ending home construction projects. "Look at this man," she teased him in front of us. "He's got good slave legs," she matter-of-factly declared. I looked at Manoel to see if he seemed offended with Guiomar's joking reference to one of Brazil's most shameful eras. Not at all. Instead, he merely smirked and rolled his eyes. He was used to her teasing, apparently. Guiomar continued, oblivious to my discomfort. "You know how you can tell a good slave? You know?" she insisted. I was speechless, which only gave her the opening she needed. "By the thin part down there." She pointed at Manoel's lower leg. "A good slave has a thin part below the calf, right above the ankle. This means they can work hard, they can bear lots of weight." She cackled loudly and looked in Manuel's direction again. "He'd fetch a lot of money today!" Manoel still smirked, continuing up the stairs silently as he carried Guiomar's cement on his back.

Growling Bellies

Jitilene and I have a date. We are going to a river council meeting that is supposed to map out communities' concerns about environmental management of the main regional river, the Rio de Contas, which serves as the gateway and transportation passage for nearly all of the regions' quilombos. Government agency and civil society representatives mill about the grounds of the quiet pousada, which is separated by a high wall from the noise and bustle of Itacaré. The *gente grande*, or big people, as "very important people" are

---

quilombolas' experiences allows us to begin with history and silence but also to incorporate memory and agency in quilombolas' lives as social movement actors. This also allows us to think not only about how people are constrained by the spaces and practices in and through which they are silenced but also about how they are agents in the production as well as the breaking of these silences.

20. In one of the most sophisticated ethnographies of Brazil in the past few decades, Goldstein reveals how "humor is one way of bearing witness to the tragic realities of life," an expression of discontent (2003, 16). Humor is a form of both resistance and survival.

called in Portuguese, slip away periodically to make calls on their cell phones. The quilombola leaders not only don't have cell phones—they seem not to want to miss any part of the meeting. Their participation comes at a price. At a coffee break, one of the regional leaders for a nearby quilombo laments to me about how much effort it took to get here. "I had to rise in the early-morning darkness at about four a.m., paddle a canoe, take a bus, and then walk for a few miles to attend this environmental advisory council meeting," he recounts. "It isn't easy. But at least it wasn't raining, for often it is." I also know his participation in the meeting meant forgoing a day's work on his farm, but this remains unsaid.

I sit next to Adriana, a young Afro-Brazilian woman who has just been hired by Floresta Viva to work with quilombos. She is intended to be a bridge between the NGO and the community, for she is a quilombola herself. We watch attentively as the different community representatives, including Jit-ilene, assemble at the head table to talk about the Rio de Contas. A biologist talks about how the crab population is slowly coming back after the oil spill on the river a few years ago. A government official describes a bridge that will be completed in the future, cutting the time for transportation and travel to Bahia's capital, Salvador, by hours. As the presentations begin, Adriana looks over at me and exclaims in an audible whisper, "I would be so nervous if I were Jitilene! She must have 'a chill in her belly'! I can't imagine having to talk in front of all of these people." I think of Jitilene and how she is still in high school yet sitting at the table with people decades older, people who own businesses, who work for the state, *gente formado*, as Guiomar says. Maybe Adriana has a point.[21]

One by one, the representatives of the various communities along the Rio de Contas introduce themselves. When Jitilene's turn comes, she speaks softly but articulately, announcing that she represents the Santa Amaro qui-lombo and lives up the Rio de Contas. After the panel discussion, as she

---

21. Brazilian anthropologist Roberto Da Matta (1979) engages in an extensive analysis of the common expression used throughout the country, "*Você sabe com quem está falando?*" or "Do you know who you are talking to?" a phrase that is often uttered in situations of conflict to insert class hierarchy and establish one person's authority over another. Goldstein notes, "Every act is mitigated through class position and is implicitly a class act" (2003, 9), while Cal-deira discusses how middle-class activists try to present meetings with working class people as "*bate-papos*," or simple "chats," when in reality the class separation between these groups makes a chat among equals outright impossible (1988, 447).

returns to her seat, I lean over to congratulate her as everyone tends to do after public speeches in Brazil. She rolls her eyes and utters, almost inaudibly, "Water and forest, water and forest—no one ever speaks of my hunger." Amid countless lectures on river ecology and government-led infrastructure initiatives, the most pressing issue faced by quilombo communities—people's hunger—is erased in the official discourse of public meetings.

Dona Otília has schooled me on how quilombo landownership and hunger are often interconnected. The Santa Amaro quilombo, for example, is sandwiched between the properties of a large landowner and the banks of the river and the small territory that exists officially as the quilombo provides neither adequate land nor soil for farming of a scale that can support the community. Once, when we visited Santa Amaro, Otília had remarked of the quilombolas residing there, "Their food is in the mangrove—when it provides. When it doesn't provide, they go hungry." I also remember how, when I had attended meetings at Santa Amaro with Floresta Viva, the first act before any business began was to prepare a community meal. Only after that, once people's hunger had been quelled, could the reason for the visit come to the fore.[22]

The realities of hunger are present in the literature, the music, the popular culture of Brazil's Northeast.[23] These realities are also very much present in Southern Bahia. Though no one talks openly of hunger, sometimes Guiomar points out people in town and whispers that they come to her house "when they don't have a piece of bread to eat." I know that one of these people is the Master. Guiomar and I walk past his house, located in the wealthiest

---

22. A common practice is for the NGO to bring everything needed for a substantial lunch for the entire community. Upon arrival, the elderly quilombola women immediately begin the long process of cooking a meal for those gathered that day. It takes several hours, but this quells people's literal hunger and provides an opportunity for them to come together and chat informally. It also creates a captive venue in which they can undertake the necessary bureaucratic work for maintaining the quilombo association, such as signing documents, hearing about upcoming training opportunities, and the like. Hunger is never openly discussed, though it is present. In one instance, while an NGO was delivering a training session that took longer than the people had intended, one of the participants grew visibly irritated. "I need to fish," she stated, simply conveying that to eat later that day or the next, she needed to catch her food.

23. Nancy Scheper-Hughes describes how hunger pervades the lives of the poor, particularly in Brazil's Northeast: "The hunger of the *zona da mata* [forested lands] is constant and chronic. . . . It is the hunger of those who eat every day but of insufficient quantity, or of an inferior quality, or an impoverished variety, which leaves them dissatisfied and hungry" (1992, 137).

Figure 13 Lunch at the quilombo before official business begins.

neighborhood of Itacaré, which sprouted up just two decades ago in response to the growing tourism industry. Chic pousadas line narrow dirt roads riddled with potholes. "This place is a disgrace," she scoffs as we walk gingerly to avoid splashing mud on our legs. As we pass the Master's house, she nods and remarks, "He's a stubborn old man." I look up at his humble but massive structure of a home in a state of disrepair. The white paint seems to have been slapped on carelessly. A beach *canga*, or tapestry, with a picture of Bob Marley covers the large front window. "He sleeps right there," Guiomar says, nodding at the window. A series of open doorways line a narrow walkway, which leads to an open-air kitchen in back. Master Jamaica often proudly declares that he chooses to cook over an open fire. "The food is best this way," he states. And here, around the fire, is where his "tribe," meaning his capoeira students, often gather on a Sunday to play music, cook rice and beans and sausages, and drink *cachaça*, Brazilian rum. "Beer makes you fat," these sinewy athletes claim.

The upper level of the house is open and unfinished, as if someone started constructing it and then ran out of money, or changed their mind, or perhaps both. I think back to when construction on this second level began with the "support," meaning the funds, of a young woman from Norway who had fallen in love with capoeira as well as with the Master himself. For a time, he held capoeira lessons on the second floor every afternoon and evening. But this was back in the heyday of the tribe, when Neto and Comprido and Manso were all here. When the tribe held regular *rodas,* or capoeira circles, not just for tourists but for themselves, to practice and celebrate and improve their art.

Now the Norwegian woman is long gone, and the capoeiristas have moved their art to Europe. The Master, too, is nearly gone himself, which is what makes Guiomar remark as we continue past the house, "He is dying. You know, he has diabetes and he doesn't take care of himself. He doesn't eat well. People in the town have started to bring him food." I nod quietly, hoping he is not lying on his bed just beyond the Bob Marley blanket, listening to this exchange. Voices travel far in the night here. I know that the Master lives in poverty, despite being one of the most iconic cultural figures in the town as well as the entire region.

But hunger is not openly discussed, not on the streets of Itacaré nor on the official agenda of public meetings like the Rio de Contas environmental council event. Jitilene has sometimes confessed to me, "People are going hungry," when we talk of her quilombo community. But she does not speak openly of this when she attends an event as a formal representative of the quilombo communities. Is hunger erased in public contexts because of its connection to poverty, shame, and oppression? Or perhaps those who represent quilombo communities in Southern Bahia or those who serve as capoeira masters refrain from raising the realities of their hunger lest this run counter to the identity of strength they are trying to portray? I don't know for sure, but I do know that just as connections to slavery are ephemeral and hard to pinpoint and just as contemporary racial oppression is silenced by Brazil's timeless myth of racial democracy, hunger and poverty are also silent yet pervasive realities in this place.

I also know that hunger is not only physical; it is also metaphorical. It describes how quilombo communities are marginalized amid the processes meant to include them. It describes how quilombolas are hungry for promises to be kept and for their realities to be considered in attempts to rewrite their past trajectory in the present. Jitilene's mother, Dona Julia, has lived this reality for nearly eight decades. Because Santa Amaro is one of the easiest quilombo communities to access in Southern Bahia, it is often visited not only by NGOs but also by international donors and state agencies interested in working with quilombo communities. One morning when I am visiting Dona Julia's booth at Itacaré's Sunday fair where she sells cachaça and freshly baked bread, I ask her who has been by Santa Amaro lately, referring to the influx of civil society and government organizations eager to set up projects. Dona Julia laughs ruefully, sighs, and says, "Oh, these people come and go, making many promises to the community that are quickly forgotten. When they leave, we think, '*Mais um*,' one more." Another quilombo leader, Cosme, has come by looking for a cup of coffee. Overhearing our conversation, he offers his two cents: "At times they bring people from outside to quilombos to get to know the quilombo, to bring projects, and then it becomes a typical project with an NGO, without anyone responsible that you can question about it." He nods his round face earnestly and wipes a bead of sweat. Dona Julia utters, "Pode creer," you better believe it.

I, too, have witnessed countless failed development efforts in attempts to make something happen in places that desperately need to reverse centuries of behavior. Recently, I had helped the staff of a regional NGO load 100 chicks for Santa Amaro into small dugout canoes. It was a chaotic endeavor, to say the least. A few peeping chicks that had escaped the box they were contained in ran about the floor of one canoe as we dodged them with our feet and tried to keep the canoe, which was very low in the water, not only upright but moving forward. As we arrived at the quilombo, we were met by the community members, who were eager to unload the precious cargo. A line of chicken-carrying folks paraded to a large coop, big enough for about three dozen people to gather in. There, folks listened carefully to the instructions of the technical expert, who was trained in animal husbandry, as he patiently explained how the chickens were to be fed, watered, cared for. Half of the chickens were to be used for egg production, while the others were to be raised as *galinha caipira*. Everyone, especially tourists, loves galinha caipira. The boxes of egg and meat chickens were separated, and the NGO technician began taking the tiny birds out of their boxes, gently placing them on the

straw. In this manner, what I call "the chicken project," intended to be repli-
cated at other quilombo communities throughout the region, was launched.

I was initially, and naïvely, impressed. Safely transporting 100 chicks in a
shallow dugout canoe was no small task. Plus, the community had spent days
constructing the best chicken coop I had ever seen. But Dona Otília, as usual,
had a less-than-optimistic view. "That's bullshit," she said when I told her of
the chicken project. She surmised the chickens would either die or run away.

As countless tales of failed development projects have shown, outsiders
entering a community that has long been marginalized often raise hopes that
things will change. The numerous projects offered to quilombos throughout
the region—beekeeping, chicken raising, artisanal craft production for sale
to tourists, making organic fertilizer, sewing, and guide training—sometimes
work, and sometimes don't. In this way, quilombo communities are remembered, yet sometimes forgotten, by the agents of development here in Southern Bahia. They are hungry for these development opportunities not just to

Figure 14 The beginning of the famous chicken project—
the arrival of 100 chicks in canoes.

be implemented but also to be tailored to their particular historical, political, economic, and social contexts. As NGOs, state agencies, filmmakers, authors, and musicians increasingly turn their attention to quilombo communities, the communities themselves are many times put in a position that is the opposite of empowerment. They are often kept waiting—waiting to be remembered, waiting to be heard. The questions surrounding these alliances transcend the specificity of projects like the chicken endeavor and raise questions of how power is demonstrated, contested, and perhaps even shifted as contemporary quilombos vie for visibility in Southern Bahia.[24]

At the fair that day, looking for some indication of success, I ask Dona Julia how the "100 chicks" project is going. "Oh, it went well," she said. "At first." She glances away and continues in a low voice. "But then we never got the second round of chickens we were supposed to get." Her voice trails off, not saying why or where things went wrong. "But we just need to wait, I guess, right?" She looks up at me from her seat behind the coffee canister, chuckles, and adds, "We have to be patient." She looks away, in the direction of the river, where her community is situated, where people wait.

"Playing"

"Where are your pants?" Master Jamaica yells to me as I come into the capoeira performance space, the crowded back of a bar on the main street. The tourists all stop and glance over at me, wondering what the heck he is talking about. I chuckle, for he is referring to my capoeira pants. Now that

---

24. This interest in quilombo communities is due both to their growing visibility in Brazil's social movement spectrum and to the reality that quilombos are some of the poorest communities in Southern Bahia, which also ties to the frequent ambiguity of their landownership and the precarity of their livelihoods. For example, in Southern Bahia, many quilombos have a historical practice of growing manioc, which dates back to the 1700s (Reis 1996). Shifting cultivation of manioc, however, is no longer allowed today, as the quilombos are located in an APA, and slash-and-burn practices go against APA regulations. Recognizing this, civil society and state actors are moving in to develop new livelihood possibilities for people who have long been marginalized. This comes with a responsibility, however. Cosme, the president of a quilombo association in the region, notes that quilombo communities can either work on their own or with an NGO. But, he states, "if communities choose to work with an NGO, the NGO has to be responsible for all it brings to the community. Without this responsibility, *ninguém vai para canto nenhum,* or no one progresses at all."

I have started taking capoeira classes, it seems I have a responsibility to show up, to play capoeira publicly despite the fact that I am terrible at it, to help to perpetuate this dance, this fight, this memory and present-day reality all wrapped into one. I stammer something about how I've just returned from the field and I'll play next time. Jamaica scoffs and looks away, visibly annoyed.

But while Max is too shy, I know Ella will join in the roda with or without her capoeira pants. She has been baptized Borboleta, Butterfly, by the Master while I am called Olhos da Águia, Eyes of the Eagle. Both of us have been "practicing" capoeira, as they say, for some time now. Each day, Max and Ella take classes with Guiomar's ex-husband, Edmundo, who has come to live with his new wife in town for a bit. The children move easily, they sing, they play the gourd, the drums. In many ways, they melt into what this place was, and what it is, more fluidly than I do.

My capoeira is played with a group of middle-aged women—"the capoeira moms," as I think of us. We try to tumble and cartwheel. We sing the songs and play the music that is foundational to the practice. And we all listen obediently to Master Jamaica, who shames us in a way no one else can, yet still we come back for more. It is part of the play.

I continually marvel at how the presence of capoeira pervades this place. It is the *alma*, the soul, of the region and not just for tourists but, even more importantly, for those who live here. Each night young men, and a few women, walk through the streets wearing their white bell-bottomed pants on their way to the roda. Jamaica is the best-known regional master, for he is the most overtly public, which brings him criticism. But there are others, too. Master Paulo, who teaches a few blocks off the main street. Master Edmundo, Guiomar's ex. And Masters Cabello and Tisza, in the nearby town of Serra Grande. These men, and a single woman, Tisza, are a strong presence both individually and collectively. Some are quiet teachers, practicing their craft and teaching others who will come after them. Others, like Jamaica, are marked by their loud brashness, their music, their wild gyrations of capoeira that call out to be noticed, to be respected. And as the capoeiristas spin through the air and sing, they collectively break the silences of the past that permeate this place.

Figure 15 Master Jamaica doing capoeira with his daughter and mine.

Documenting Things That Matter

While just a few years ago it seemed that no one talked about the quilombos that were located "up the river a ways" or "over there, in the interior" apart from the well-known coastal tourist towns in Southern Bahia,[25] now these communities are highlighted in scholarly studies, development efforts, conservation initiatives, government projects, and even films.[26] Chuckling, Jit-

---

25. Decades ago, anthropologist Anna Tsing (1993) called for ethnographies of marginalization in "out of the way places." At a cursory glance, quilombos epitomize this description. Though Itacaré's neighborhood of Porto de Trás is an urban quilombo, the other regional quilombos up the Rio de Contas can be somewhat difficult to reach. One can take a ferry with a car, and then drive for more than an hour on a rough dirt road, not always passable after heavy rains. Another option is to find a canoe, paddle up the Rio de Contas for thirty minutes, then navigate through narrow mangrove river tributaries. More distant quilombos require hours of travel on the river or hiking along inland trails. Many cannot be accessed for days if intense seasonal rains make the river dangerous to navigate or roads impassable.

26. Brazilian and Brazilianist anthropologists agree. Carlos Diegues produced the film *Quilombo* in 1984; since then, interest in these communities has grown vastly. In 2003, *Quilombo Country* was produced by Leonard Abrams, and there have been successful *telenovelas*, soap operas, about slavery called *Xica da Silva* and *Isaura the Slave* and *Time Doesn't Stop*. Richard Price observes, "'Quilombo' has taken a prominent place in Brazil's rich forest sym-

ilene wryly claims, "Now we're right up there next to the Indigenous!"[27] I find myself wondering if and how quilombo communities and quilombola people can emerge from centuries of social and political neglect. I need to know more.

Jitilene and I are finally sitting down for a formal interview. We meet in a clean, sparsely furnished neighborhood bar attached to her house. The bar is run by her mother, Dona Julia. One bare fluorescent light bulb hangs in the center of the room, which is painted a cheery yellow. Two small white, unstable plastic tables, which are normally reserved for simple street-side bars, are lined up against one wall, and a large refrigerator housing cold beer stands in the back of the long, narrow room. She waits for me to begin the conversation. I am hot, and it is late, so I tell her I'll buy us some drinks. A beer for me, a Coke for her. As usual, Jitilene is shy at first. We crack open our beverages, and I start to explain why I want to speak pointedly with her. She nods and brusquely interrupts me as I stumble through my standard introduction on how I am studying social movements and she is a key leader, and instead flatly states, "We need a camera." Caught off guard, I try to gently deflect the request by suggesting she approach Floresta Viva to perhaps loan her a camera. I know the NGO has several; I had even brought one for Sayonara to document quilombolas' lived experiences. Jitilene will have none of this. "Nope, we need our own," she emphasizes. I try to understand this seemingly urgent need. "*Why* do you need a camera?" I inquire.

"It's complicated," she starts. Then stops. Then starts again. "You see, there's this state foundation that wants to give us funds to help us do repairs on the homes on the quilombo. But our quilombo association has to document what the money will be spent on if we're going to apply for a grant to get these funds. So we need a camera." What she doesn't mention, but I know, is that they also need money to print photos, a computer to write a project, training in using the computer, and an incredible amount of tenacity to

---

bols" (1998, 246), and more recently, Ilka Boaventura Leite claims, "In contemporary Brazil, 'quilombo' is a term that signifies transformation. It is heard everywhere, from popular demonstrations to political affairs of state" (2012, 250).

27. Leite (2000, 333) points out that many of the territorial struggles are similar between Indigenous Brazilians and quilombolas. Nevertheless, Brazilian Indigenous populations arguably have much more public visibility than quilombolas.

navigate the long and bureaucratic processes of being remembered amid the social structures that forgot these people long ago.[28]

But the camera is an important first step for her, as the president of the quilombo, being able to visualize and present the realities of where she lives through her own gaze rather than one mediated by an NGO or government foundation. She is bridging the past and the present. Jitilene and her community don't need a camera for performing quilombo identity or representing themselves as romanticized versions of Brazil's African heritage. The quilombo needs the camera instrumentally, as a tool for documenting their needs in the present. The quilombo needs to take hold of this story, this reality, itself. Plus, I remember how Jitilene had once remarked to me, "People come to hear our story, *para pesquisar agente*, to study us, and then they go away and we never hear from them again." Mais um, as her mother had said. Maybe a camera would allow *them* to be the agents, rather than the objects, of their own research.[29]

Recognizing Presence

I drive the erratically trustworthy fuscinha to pick up Dona Otília for one of our regular journeys to Ilhéus. She is late, and I am hot and annoyed. But she is Dona Otília, and no one can speak crossly to her—though she has the power to do so to everyone she comes across if she pleases. Pulling into her driveway, I call to her. She walks toward the car. I open the door to let her in when abruptly she turns and, without a word of explanation, walks back toward her house. There is nothing to do but wait. Perhaps I have offended her somehow, I think. After a few minutes, however, she returns bearing a large bound book, which she opens. Still with no explanation, she begins

---

28. The institute in charge of restoring quilombo homes was the Bahian Institute of Artistic and Cultural Patrimony, also one of the key players in the restoration of the Pelourinho neighborhood in Salvador. See Collins (2008) for a study of how Bahia's colonial past is employed to create present-day opportunities for domination, as well as resistance, of people who both adhere to as well as stand apart from categories of cultural patrimony.

29. For example, Adams et al. (2013) argue that conservation and development policies (related to Brazil's Forest Code regulations) that affect quilombo communities in the Atlantic Forest should be less restrictive and allow for new opportunities for organization and innovation within quilombos.

reading aloud the names of four quilombos in Southern Bahia. She reads all of the text—names, procedural numbers, locations, and dates—and then closes the heavy book with a bang. I try to figure out what is going on. Then she simply states, "These four quilombos received official federal designation. Three other quilombos are still unrecognized."

Otília's pride in this formal recognition is apparent. But what is the significance of this registry *em branco*, on paper, as they say in Brazil? While the written names in the book give these communities official recognition, we both know they mask the ways in which people are forgotten amid the very structures that are established to remember them; while legalizing land tenure and appearing in registries are fundamental first steps, if this can't be enforced in practice, it makes little difference in people's lives. But in her abrupt way, I know that Dona Otília is making an important point. She is showing me that official designation is a critical first step. She is pointing out that recognition matters despite the imperfections of the broader structural context within which it exists.[30]

Remembering, Recovering, "Reaching for Resgate"

There are words in Portuguese that don't exist in English, and one of these is *resgate*, akin to "cultural rescue, remembering, taking back"—all summed up in one. Resgate focuses on the memories and stories quilombolas have passed down from generation to generation. Increasingly, younger quilombola leaders are starting to talk about remembering, recording, and capturing memories

---

30. The negotiations these communities had navigated to obtain state recognition marked a political struggle in which historically "oppressed classes" were finally beginning to uncover their long "suppressed histories" (Benjamin 1969). French (2002) has argued that obtaining official designation is a process in identity formation and that becoming recognized as a quilombo is a necessary step on the trajectory to being remembered. See also Shore (2017, 58), who describes the glacially slow processes of recognition in São Paulo's Atlantic Forest region and how quilombolas engage in place-based socio-spatial struggles.

Perry (2016) talks about how intersectionality can be strategically mobilized by Afro-Brazilian women activists to bring about social and political change, and Farfán-Santos (2015, 2016) explores the lacuna between legal recognition and social and political rights recognition, as well as the ways that Afro-Brazilian activists are mobilizing to lessen this gap. Sean Mitchell's (2017) brilliant ethnography reveals how inequalities are produced and reproduced among quilombo communities due to the presence of the Alcântara military base in Northeastern Brazil.

in the present—before it is too late.[31] One of these people is Comprido, a young capoerista who has been working to make his life here. Unlike Neto, Guiomar's son who went away to Sweden, and numerous other capoeiristas who followed suit and now live scattered about Europe, Comprido chose to stay. His mom lives here. He has a house. He has history. Perhaps he also has no choice.

One afternoon I trudge up the cobblestoned street in search of his home deep in the heart of the Porto de Trás neighborhood, which has one main street with a few narrow pathways jutting off. I know that if I ask around, I'll quickly find the right structure. Everyone knows everyone here. Sure enough, the first folks I see, two men having a midday stoop-side chat, cheerfully direct me toward it. I wind through a labyrinth of small walkways lined with well-cared-for tropical plants in bright pots next to doorways. The roads here are designed for foot traffic only; few have earned a driver's license, and even fewer have a car. I gaze in the open doors as I pass, each home revealing people's business and goings-on. I see people lounging after lunch, taking a midday rest on tattered couches. A woman grooms her daughter's hair on the front stoop, carefully capturing it into neat braids while the little girl sits completely still, looking somewhat pained. Others cook, the sizzling of beans emanating from their open doorways. I stand at the bottom of a staircase leading to what I think is Comprido's home, and my heart sinks. The closed door and windows indicate that he has forgotten about our appointment or gotten a last-minute guiding gig with tourists. Paying work is something that is not to be turned down. In Bahia, both forgetting or finding something better to do are logical reasons for a last-minute change of plans. People don't get upset about things like this, and over time I have learned to accept the fluidity of time, appointments, and promises.

Just in case, I give a call and a couple of loud claps to ensure he isn't home. To my surprise, I hear some scuffling, and Comprido opens the door. He has

---

31. In this, they are scrambling to record particular histories and memories of quilombolas in an effort to promote a social and collective remembering that is initiated and controlled by quilombolas themselves. Anthropologist Paul Connerton argues that when subordinate groups are given control of these processes of articulation, "not only will most of the details be different, but . . . the very construction of meaningful shapes will obey a different principle" (1989, 19). By building alliances, choosing their representatives, and advocating for the recognition of their dynamic and hybrid identities, quilombolas are finding ways to exert their voice amid historical and present-day challenges.

been watching television, relaxing from the midafternoon heat and waiting for me. As he quickly takes off his glasses, it suddenly occurs to me that I have never seen him wearing them. While glasses are hardly uncommon, what strikes me is how they seem so very out of place on Comprido. Capoeiristas are famous for rippled muscles, for flaunting their toned arms in capoeira rodas. The look of a studious scholar comes as a surprise. Later, I learn from Comprido that the designation of Porto de Trás as a quilombo led to three youth from the community gaining a spot in the regional college by claiming their quilombola identity. I wonder to myself if and how Comprido will use this designation to advance his own opportunities in life.

"Can I borrow your audio recorder?" he begins our conversation, as I take it out of my bag. I look up quizzically, trying to understand why my audio recorder is in demand. He continues, "You see, the elders in our community are dying, and they have so many stories that need to be recorded. We need to document this stuff before they're gone." He then proceeds to name the people in the neighborhood he and his friends want to interview, a list that quickly reaches double digits. Dona Otília, a "non-quilombola" per se, is one of the first people Comprido wants to interview about the history of the region.[32] "Why her?" I inquire. "She has the most knowledge; she is an authentic leader. Plus, she will talk to us," he rationalizes, for he knows that Dona Otília is working on a book on the regional quilombo history.

Comprido also knows that eliciting stories of the past from quilombolas in the present is not as easy as merely obtaining a recorder. How do you access silences and memories? Here lies a paradox—to be quilombola, you must recognize a painful history of slavery, and as Comprido says, "These people are the most resistant of all. They, the quilombola elders, are the hardest to get the stories from!" He shrugs, gives a little laugh at the irony of this, and we continue our interview.

---

32. Quilombola conceptions of authenticity and representation can be malleable and strategic. French (2009) shows in her research with quilombo communities that sometimes these communities strategically adopt other identities, such as Afro-Brazilian or Indigenous identities, depending on particular social and political situations; racial politics can be manipulated to either highlight or downplay quilombo identity, depending on historic, economic, and social context. Leite argues that quilombola communities are not racial categories but rightful forms of social organization (2015). But what happens when quilombo identities are unknown, confused, or perhaps even undesirable to the people who comprise them?

Figure 16 Dona Otília communicating things that matter to youth in the community.

Speaking

There is another capoeira roda tonight. Our family parades into Jungle Bar, which has become the spot where these rodas take place, three times a week, at eight or so. Master Jamaica is having a good night; he is in full force. He paces like a caged wild cat, calling out to those who are watching but not taking part. I hide in the back, trying to blend into the crowd. But he notices everything, and as always, when he sees me, he yells, "Where are your pants?" In his eyes, it is inexcusable to come and merely observe. Capoeira is a living art form perpetuated by those who practice it. If it ceases to be done, it dies. Keeping this alive is what the Master has dedicated his life to. As the music crescendos, the capoeiristas become frenzied. The drumbeat gets faster. The cowbell rings louder. People lean in, singing, and those in the center of the circle start playing hard. They do flips in the air, nearly hitting each other's heads as they twist and turn. The circle grows smaller and smaller as the intensity increases when suddenly, with a start, the Master holds up his hand. The music and movement stop in a split second. "Get over here," he orders the capoerista tribal members, pointing to the ground. They scramble to take a seat in the circle, sweating profusely, trying to catch their breath while directing their eyes on him. The lesson begins.

"What is capoeira?" he calls out rhetorically, for no one has the courage to answer him in public. He is a formidable figure, fifty-five years of lean muscle, broken teeth, and a raspy voice. He walks about the small space with a commanding air, directing his attention between the capoeiristas and the public watching this spectacle and history lesson rolled into one. He exclaims again, "What is capoeira? It is freedom!" The capoeira players nod their heads, uttering, "*Isso ai*, that's right," or "Pode creer, you can believe it." Someone whistles loudly; a few others clap their hands. For the next fifteen minutes, the Master proceeds to deliver a lecture that blends history, culture, music, dance, and rebellion. He talks about fleeing the bonds of slavery in the dead of night. He talks about hiding in the capoeira grasses, making the connection between escape and rebellion and the physical, oral, and musical practice they are engaging in today to conjure this history in the present. After the lecture, directed not as much at his tribe, who knows this well, but more at the observing tourists who hail from São Paulo, Rio de Janeiro, perhaps even Europe, they close the night with a final round of movement, the biggest and fastest and fiercest yet. The audible drumbeat vibrates through the streets of the small town, breaking the silence and signaling that people are coming together to connect and embody a historical legacy with a contemporary identity.

Writing, Finding an Official Voice

Like the Master, Dona Otília, too, chooses voice. "Otília has no problem talking with people, with federal ministers, with gente grande," the former mayor of Itacaré tells me one day as we speak about the region. "She speaks her mind, no matter what!" He gives a small laugh. I am not sure what he means by this really, until one afternoon, Otília and I set off to drive an hour and a half from Itacaré to CEPLAC, the federal cacao research center near Maisa's land reform settlement. When I ask her if she has an appointment, she replies confidently, "Oh, they'll see me," and continues to look at the road ahead of us, signaling that the conversation is closed. I am somewhat doubtful, but Dona Otília is not someone you question, nor is she someone you throw out of your car mid-trip. I bring Maya and Guiomar with me. They are on one mission, Otília and I on another. We arrive and find our way to their destination first, for I cannot do my work until they are both settled.

Maya beams happily when we arrive—a small house with more than three dozen sloths in an open-air caged repository. "The Sloth Lady" who takes care of the creatures is happy to see us; she knows us by now, as this place is sanctuary not only for sloths but also for my children. It is a place where they can come into the shade and quiet of the forest and while away the afternoon playing with these gentle creatures that have been rescued and are in various states of recuperation. Guiomar gives us a wave and a nod, and we are off to our meeting.

We arrive at the entrance to the institute, park the car, and walk down the freshly mopped vacant hall. Coming to the reception area, we are informed that the director is in a meeting. I look at Otília, refraining from saying, "I told you so!" as I know this could take hours. But then the receptionist returns, noting we should wait. Sure enough, before long the director emerges, gives Dona Otília a warm hug, and invites us into his office for a personal chat about how they need a university closer to Itacaré, about how farmers in the region are faring, about his family, about her family. After an hour, we leave with a plan for mobilizing a federal governmental institution to address the region's educational needs as well as with special handmade presents of premium-quality chocolate made on the premises of CEPLAC. As we walk down the hall, the director pauses at a black-and-white photo montage that lines the wall. "Wait, where are you?" he asks Otília, staring intently at the people in the pictures. In a few seconds, he locates her. She is much younger, surrounded by all men, somewhere on a highway in Bahia. "Oh yeah, that was my activist days," she states roughly, and quickly heads down the hall with her characteristic deliberate waddle. I have a sense that this is how she gets things done. She shows up, and speaks up, and makes things happen. She runs after. And her efforts don't stop at the borders of Bahia.

Barack Obama has just been elected President of the United States. For Dona Otília, this is a big deal. I had no idea that she paid much attention to international news, but a few weeks earlier she had approached me and brusquely asked, "Hey, you're from America, right?" "Yes," I said hesitantly, as sometimes in Brazil being from the United States is more often reviled than revered. "I need to reach your president," she continued, matter-of-factly. "I have some things to say, so I'm writing to him."

And so, one blistering afternoon I find myself sitting with Otília in her sparse gathering space, watching the cars reel into the speed bump next to her property at eighty kilometers an hour, slow to a crawl to surpass the bump, and then accelerate rapidly again. Her house is a short walk beyond, still close to the highway but not close enough to see it. When I first met Otília, her space housed a popular restaurant. Later, with funding from elsewhere, the state or an NGO, it was turned into a community gathering area for large public meetings. The roof was covered in artisanal palms, making it look quaint, though the location adjacent to the road slightly challenged this ambiance. For a time, we had many meetings here. Different associations and groups would gather on wooden benches and hook up a sound system to hear the speakers above the sounds of passing cars or, sometimes, the pounding rain. But each year, the space fell into a more advanced state of disarray. Without a constant flow of funding, the roof started leaking and holes emerged in the cement floor. Now the benches are stacked on the side of the space, it has ceased to be a vibrant place of communal gathering. But Otília still sits here, using the space, making things happen.

Otília and I pull two chairs around a single table. We are women on a mission; we have a short time to translate a ten-page poem she wrote for President Obama. Max runs about taking pictures as we get to work. "Why do you want to do this?" I question her as she hands me page after page of her careful writing, reading it aloud to make sure I capture every word. "Well, he's the first Black president of your country, right?" she asks rhetorically. "Here in Brazil, in Bahia, though we have quilombos and quilombolas, we don't know how to speak out. We need to find our voices. So this is what I'm doing, giving voice to this region, to the people who have a history here. What else can I do but try?"

Where this will go, I have no idea. But the following year, when I tour a friend's newly built house in Colorado, I suddenly pause in her foyer. "Is that a picture of you with Michelle Obama?" I ask incredulously. "Yeah," she pauses. "Remember we are both from Chicago?" "I have a somewhat unusual job for you," I begin.

To this day, I'm not sure if the poem got into Michelle Obama's hands, or into President Obama's, for that matter. Perhaps its purpose was merely to disperse Dona Otília's historical knowledge and activist energy out into the world. Kind of like the river dolphins, gestures here can be intense and earnest, as well as fleeting. They also have a strange way of connecting here to there, wherever there is. It might be Europe, where many of the capoeiristas now reside, or it might be the United States that is reached by an elderly woman spending her time composing a lengthy poem to the

Figure 17 Dona Otília advocating publicly for the region and all who live here.

president. Sometimes things serendipitously work: the dolphin rises right at the moment you are looking at exactly that spot in the river. Other times, you miss it. But the efforts to uncover, to recognize, that which exists here and to try to link this to that which is la fora, out there, cannot stop, no matter how difficult and fleeting they may be.

As we finish our work for the day, Dona Otília looks at me resolutely and instructs, "Put my cell phone number at the end of the poem in case he wants to reach me. I'll be here, waiting."

A POEM FROM AN ELDERLY COUNTRY WOMAN TO AN ILLUSTRIOUS PRESIDENT[33]

I ask your apologies
For my enormous boldness
To be from another continent
To live in another nation
But with the greatest humility
Hearing all of the reports
I took up my pen
And wrote in prose and verse
To pay homage to a man
Who arose among millions
With good proposals and actions
Determination and love
Tenacity and modesty
To win against discrimination.

I am proud of the Black race
With a large love and admiration
And also of the Indians
I am a descendant of these two
And for this I have suffered much
I had no way to study
In a country with much discrimination
Living the need to fill
To cultivate the land
And by this land I was sustained
But I had faith in God and courage
To confront prejudice
So I see the world in another manner
All that passes is not a mirage.

I live in a country without terror

---

33. In Portuguese, the poem adheres to Dona Otília's famous way of writing in rhyming stanzas, though this is lost in the English translation.

The worst thing here are the drugs
Here there are no volcanoes
Some small earthquakes
We don't have cyclones or tidal waves
In the south of the country some small
Hurricanes are growing
The robber here walks free
Which is terrible
Because the world is this way
We have high levels of criminality
We have a lot of work to do in this area
For people to learn
Because we don't lack conflict.

We have an excellent government
Our dear Lula
Who ascended the ramp of the Planalto
Remembering the least favored
He is a great man
In this way he diminished hunger
And is going to diminish violence
He is improving education
Investing in culture
Inspiring sports
For this I am happy
I am living in a paradise
The pride of my country.

I've seen many reports
Coming from North America
I know that inequality is great
And prejudice is strong
But you are brought from God
You have conquered the masses
We are certain of your victory
I know that you're going to help your race
They have as much suffering as we do
With the disposition to conquer

To have a Black man in power
Is the glory of the future
Better days are going to come
And we're going to thank God.

Obama, you are a child of God
I see the world helping your nation
Against the viruses contaminating the world
You are dedicated to fighting
I know that you are going to do away with problems
With your valiant determination
To fight against pollution and terror
Emitting gasses without limit
Contaminating the world
Without our rulers feeling them
They don't stop polluting
From one side terror reigns
And from the other, pollution pressures
If it doesn't stop, the earth is going to explode.

Global warming is the fruit
Of much pollution
We need to reforest the world
To have less water pollution
To stop burning and fires
The earth has its limit
Listen to what I'm going to write
Rulers aren't worried
They only think in nuclear energy
Without light they live without seeing pollution
Or feign that they don't understand
They live without a plan
Only living to earn money
With the world destroyed, where will they spend it?

But since you are going to light up the world
I am writing this to you
I know that you are going to shine on the two continents

That you are going to help our Lula succeed
Lula will shine in the South
And you will shine in the North
United, you will be strong
Holding the hand of God
To stop this hurtful war
For a world with so much but still lacks bread
Where children cry tears of hunger
In the hope of a little food
Seeing their parents lose their lives
Losing relatives, friends, and homes.

When I see you on the television
I am filled with hope
I can't know you personally
Because the distance is great
I am poor and have no way to reach you
But through a friend
Who is very dear to me
I know that this will arrive in your hands
And that you'll understand
How much we admire you
And also our worry
About this brutal and wrong system
Where all live suffocating
From so much pollution.

Barack Obama, go forward
Life has chosen you with love
It has given you this great mission
With angels on high
Holding your hands
Preparing your country
As a leader of the world
And the world is going to vibrate
With your brilliant ideas
You'll dialogue in the United Nations

Your proposals will have consistency
With strength that isn't dictatorial
You are like a bird spreading good ideas
Through your dialogue and composure.

God and the people are with you
Go forth and don't look back
Fly on the horizon
Because you are capable
And the world needs you
Saints and angels will help you
We are very worried
About terror and pollution
That suffocates the whole world
Though here we have less of this
We are worried for our brothers
The worst here are the capital cities
The rest of the country lives in peace.

I live in Itacaré, Bahia
A small town encircled by forest
With more than fifty waterways and rivers
Many waterfalls and cascades
The sixth corridor of the planet
With abundant fauna
The beaches are famous
Covered with forest and clear water
With little pollution
The greatest biodiversity of the world
I invite you to visit
This splendid beauty
Created by Mother Nature
I know that you will like it here.

I am not dictating anything
But merely expressing my worry
I am an old country woman

Who lacks a good education
But I see that you have a future
You are going to weigh anchor of the boat of Victory
Don't worry about the darkness of night
The day is going to shine like a ray of light on the horizon
Illuminating your path
Always in the arms of the people
Fly on an ocean that is calm and serene
Give your cry of peace and love
For you have protection
Who is holding your hand
Is certainly Jesus of Nazareth.

Obama, God inspired you
As he has inspired my children
My people, country, and forests
For the truth won't lose its brilliance
Through your humble and profound gestures
A smile that conquers the world
I wish you all the best
You are a young hero
I know that you will say no to war
You will bring peace to your land
For people believe in you
Like the iris waits for light
We wait for Jesus
To see you in power.

The angels on high
Are uniting to fight
For the people of Iraq
Who are crying out
With a terrible terror
With so much blood lost
Death everywhere
We feel it in our souls
And we can't complain

Because our country is another
May God protect your daughters
And illuminate your people
That, united, you will change the world
And help your family.

Obama, I take leave
Pardon me for my audacity
I am inspired here in the forest
And don't participate in your daily life
I am inspired by flowers in bloom
By the sweet taste of literature
By hearing the songs of birds
And seeing the purity of flowers
The magnitude of President Lula
Are fonts of inspiration
Steps toward peace and love
Hear my greetings
Accept an embrace and your brothers and sisters
And the reverence we owe to God on high.

Dona Otília Maisa Nogueira
Vila Marambaia
Itacaré, Bahia, Brasil
Cellular: 73-9953-5923[34]

---

34. This is a fictitious cell phone number.

Finding Present-Day Voice amid Historical Silence

How do historical forms of oppression continue in the present through silences, erasures, and new forms of marginalization, and how, in turn, are rising social movement actors working to assert their voice and power amid these dynamics today?

Quilombos are growing in their visibility throughout Brazil today as NGOs, government agencies, foundations, and academics work with these communities. With their increasing prominence in civil society and state circles, however, quilombolas confront a paradox: to legitimize and strengthen their social movement status and be recognized in the present, they must conjure memories, evidence, and identities of the past. Furthermore, quilombo communities can still be forgotten within the very structures and relations intended to give them voice and visibility in the present. Quilombolas are often still profoundly constrained by historically embedded inequities, which curtail their efforts toward recognition and development as individuals, as communities, and also as a broader, integrated movement. This chapter explores the "spaces of silence" that exist within the structures and relationships that circumscribe quilombo communities today. Uncovering silences, literal, figurative, and even structural—as well as efforts to break these silences—can help us better understand the broader trajectory of social movements toward success or failure.

Creating a Politics of Place

"We have an obligation to defend this place for the people who live here. . . . We are privileged to be born here."

—Comprido, Porto de Trás, Itacaré

Being Nativo

Guiomar and I lounge on one of Itacaré's idyllic beaches, a small cove of white sand surrounded by a half-moon of palm trees. She looks around at the lifeguards playing soccer with makeshift goals of sticks, at the children scouring tide pools for sea creatures, at the capoeiristas practicing agile moves on the sand, and observes, chuckling, "The only ones who appreciate this place more than the tourists are the nativos!"

When meeting someone in Southern Bahia, one of the first things you might hear is "*Sou nativo,*" I'm a native, a term that simply signifies "I'm of this place, born here." Being nativo holds an infinite possibility of meanings. Nativos, or locals, are the best tour guides, capoeiristas, forró teachers, and cooks. This is also strategically exploited. There is the Nativo Crêperie, the Nativo Trip and Tourism center, the Nativo Bakery. Guides who walk the beaches in search of tourists pass out business cards that clearly identify them as not just any guides but *guias nativos*, nativo guides. Sometimes, people wear shirts that proudly declare their identity in a single word: "Nativo."[1]

---

1. In one sense, nativos are difficult to categorize because in Southern Bahia, the term *nativo* applies to most people except the relatively recent transplants from towns in Bahia's interior or elsewhere in Brazil. Nativos are many of the people throughout this region, the fam-

But there can be more behind this term. My friend Catu explains this to me one morning as we sit in front of a fishing association. We watch nativos mend nets as nativo-guided Land Rovers pass us loaded with tourists for day trips to nearby beaches. He says, "When you arrive here, if you have personality, you are going to say what you think. And they, the nativos, are going to say back to you, 'You don't know anything . . . *I'm* a nativo!' The city is dominated by this attitude, especially for people who come here, but the nativos are restricted by this." He pauses, thinks for a moment, then continues, "But for me, a nativo is a person who wants good for the city. I was born here because I was born here. But I want to be able to arrive in the United States or in Japan and be treated the same way. I am a creature of the planet."

I think about a recent chat with Comprido, the capoerista who decided to stay in the region while the rest of the capoeira tribe went away to the South of Brazil or Europe. We had sprawled out on the concrete slab of the cultural center where people warmed up for capoeira classes. High on the hill in Comprido's neighborhood, Porto de Trás, the urban quilombo, the two of us looked out over the river far below and reminisced about our friends in common. Neto, Nino, and Manso were in Sweden, Norway, and the Czech Republic, respectively. They were all working somehow, moving furniture and teaching capoeira, or flipping burgers and teaching capoeira, or working as a nanny and teaching capoeira. Various forms of striving, running after, to make a living while holding on to who they are. Manso was starting his own capoeira school, Comprido relayed, which was a true marker of success. "His wife is helping him a lot," which I knew was key. No one arrives in Europe and sets up a capoeira school easily, especially when you spend your precious savings on the expensive ticket just to get there. We had also talked about Comprido's sister Zety Sá, who had met a European a while back, moved away, had a couple of kids, and now returns to Bahia once a year with family

---

ilies walking along the road with wheelbarrows of downed wood, fishers mending their nets for the next day, young guides at the bus station with pamphlets in hand eager to recruit incoming tourists for a day of visiting the region's waterfalls. They are also government leaders, staff of NGOs, doctors, and owners of small businesses. Many nativos are keenly aware of their place in the broader context of a rapidly developing Bahia and are, rightly so, increasingly positioning themselves as having historical knowledge and skills that come from "being of here." Just as when people say "I'm an environmentalist," saying "I'm nativo" can also communicate a web of significance and meaning (Geertz 1973).

in tow. "It's important to let kids know where they are from, to know their roots," Comprido had said, chuckling quietly, as he tends to do.

We had gossiped about women from Europe who come to Bahia to try to make a life here, who have children with a nativo and then find that things grow difficult as kids get older and need a good education, which is hard to find here. Most end up back in their homelands across the ocean. We shared tales of the town's elite, the business owners who leave in the off-season when the tourists are gone and come back when it gets busy, when the town is filled with *gente bonita*, or beautiful people, as they say in Brazil. And as we had talked and laughed and reminisced, Comprido suddenly stopped and mused, "You know, everyone who has money can leave and go somewhere else." He gazed out at the Rio de Contas flowing into the Atlantic Ocean, the dense forest all around us, and concluded pensively, "And us nativos? We have to stay here. We have to figure out how to make a life in this place."

Figure 18 Comprido (right) and his cousin Nero in the capoeira center where they teach their culture to nativos and visitors alike.

Meeting Catu, Socorro, Erasmo, Irado

Catu is a solid, formidable presence with big, thick arms and a large fore-head marked by deep creases indelibly etched from years of living. Each time we meet, his appearance is slightly altered. One year he has long hair with shaved sides that he piles in a small knot on top of his head like some sort of Afro-Brazilian sumo wrestler. Another year he is a male Medusa with tiny braids covering his head, while other times he has a large Afro that makes his already large frame even more imposing. But his good nature is static; he is perpetually easygoing, quick to smile, willing to take time and chat.

Catu knows everyone in town and they know him. As the owner of prime real estate in the middle of Itacaré's main street, he sees and learns quickly of all that happens here. Sitting with him on the steps of his prop-erty, as he does most nights, is an invitation for interruptions. Those who pass need to say hello or give the common greeting in Bahia, a hand slap followed by a fist bump. When people ask how he is, he often utters a single word that says it all: "*beleza*," beautiful. Beautiful could be the meal he just ate, the way his day is going, or his appreciation of life in Southern Bahia. For Catu, life is beautiful. A self-employed electrician, he has worked on countless homes or businesses in Itacaré, but he also has dreams beyond, such as building an "electrical salon" to give people practical training in the profession. "I'm going to provide a quality of life to my clients," he says. "They are going to come into my salon and have a tray of tea, a selection of liquors, a couch, a television with videos, and magazines on the environment, a small library of sorts. I want to provide a good place of comfort and education, where people can learn. I'm working on this, researching the possibilities, you know." He winks and reminds me of this dream often.

Though I don't know exactly what an "electrical salon" is, with the pace of construction happening in the region today, perhaps this dream is spot on. It would allow for a level of autonomy from both nongovernmental organiza-tions and municipal agencies alike. Independence is important to Catu. One time, when a local NGO director asked if he could store a significant amount of money in his bank account, likely to evade taxes or governmental oversight, he flatly rejected him with the explanation, "I may be black and poor, but I can maintain my name, man!" And while he is ever willing to lend a helping hand to those in need, as everyone says, "Catu doesn't work for anyone."[2]

Socorro doesn't know Catu, as she lives in Ilhéus, but she, too, conducts her life according to an internal barometer of right and wrong that is guided only by her conscience. I often sense Socorro before I see her—she bears that large of a presence. Socorro most often enters a room with a commotion, a loud voice, a strong embrace, and a deep and intense gaze into the eyes of whomever she is meeting with. She is always dressed to the nines with neatly coiffed hair that changes color regularly. Her enthusiasm is effervescent, her smile broad, her laugh more of an exclamation point than anything else. Most of all, she takes pride in her ability to get engaged with whatever needs her engagement. "My name means help!" she likes to state proudly, the word coming out as "helpie," as she tries to say it in English. There couldn't be a more fitting name for a person.

If Socorro ever lacks confidence, she hides it well. Coming from one of Southern Bahia's well-known families that has raised mayors, judges, and other community leaders, Socorro has a connection to this place and its people that is palpable. She can't walk down the streets of Ilhéus without talking with dozens of people. She has three sons and an increasing number of grandchildren. If you could say that someone "owns" a town, Socorro owns Ilhéus. And she knows all about it, voraciously reading up on everything from the city's history to its contemporary literature. As Socorro often

---

2. Though nativos certainly need jobs, this statement conveys a strong sense of independence. In other words, noting that "someone works for someone" conveys one power dynamic while saying that "you don't work for anyone" conveys the opposite. Guiomar, for example, often reminds me that our situation is unique. "I won't work for anyone but you," she says, which—to a degree—places our relationship on more equal grounding and recognizes our friendship and mutual interdependence (see Collins 2008; Goldstein 2003). Stemming from this trait of nativo independence, despite (and perhaps because of) the broader region's colonial history, autonomy and equal treatment are also important to many nativos. The presence or absence of these characteristics can create an alliance or a chasm between the nativos and the non-nativos who have moved from elsewhere to make their home here for the short or long term. One pousada owner, a transplant from Rio de Janeiro, confided that the nativo independence, pride, and self-sufficiency "make these people so hard to work with." She explained, "You have to talk in a circle around the subject you want to raise with them or else you run the risk of making them mad and having them not speak to you for months or not show up for work." In this way, nativo independence can be a strategic repositioning of power against non-nativo employers who are increasingly moving to the region.

states to those she meets for the first time, "This is where I was born, where I studied and worked, where I married three times, and where I raised three children. I never want to leave. I live in the best place in the world." But perhaps the best evidence of her love for Southern Bahia is in her poetry, which conveys her sense of place, the sheer bliss she feels in living here. She writes: "It was in the middle of nature, in from the ocean, alongside the mangroves and in the cradle of the sea where I first cried out, where I took my first steps, where I learned to run, where I learned to swim without being afraid of the current, where I discovered the world, still a world very limited, but, for me, infinite. Here the sky was bluer, there were more stars, the beach and the birds were my colors."[3]

Like Catu and Socorro, Erasmo and Irado are also nativo *figuras*, characters, as people here refer to those who have a strong presence. Both of these hard-working nativos are undeterred by the lot they were dealt in life. They know they likely will never leave Southern Bahia, though they have tried. Both live countless iterations of a life within a span of about five square kilometers.

Erasmo was raised by a single mother, a woman he calls "my queen, my warrior woman." Another guerreira. He learned at a young age to strategically manipulate the tourist economy. "I was a messenger boy here in Itacaré," he says. "I would go run and buy bread for people in town. Beans for people who are vacationing here were three reais, while for me they were one." Like Catu, he, too, has undergone various transformations, which is perhaps one of the secrets of remaining relevant, of being able to shape-shift accordingly in a small town. One year he is an earnest young man trying to practice his English on every tourist he can find. He works in a store. He keeps his head down, earns money, helps his mom. The next year he is a capoeirista, grows his hair into dreadlocks, and starts to talk differently. He exudes more confidence as he runs with the "cool crowd" of capoeiristas, yet he still seems to be a bit on the periphery as he didn't grow up with this culture as the others did. Another year he cuts his dreads, gets a passport, has a going-away party, and sets sail on a cruise ship to see the world and "make real money," as he puts it. "This is my time," he declares confidently. But he comes back sooner than we all expect. No one knows why. He then opens a local bar with a friend from

---

3. This is taken from a poetic account Socorro wrote of her life, called "Me and the Indians." The region's primary Indigenous tribe is the Pataxó-Hãhãhãi (see Andrade Souza 2017 on the violent removal of the Pataxó-Hãhãhãi).

capoeira, a guy named Adriano whom the Master has dubbed Ears, never missing an opportunity to memorialize a physical trait by transforming it into an eternal capoeira name.

Then there is Irado, the constant presence in our lives in Itacaré. Though he isn't technically nativo, he is thought of as such, with his ubiquitous presence. No one knows him by any other name, nor will they ever, I imagine. *Irado* means angry, spunky, rampageous, and he is all of this and more. Rumor has it Irado was a homeless kid on the streets of Salvador and somehow found his way to Itacaré. Guiomar remembers those times. "He came by my house when he was hungry," she says, then adds, "and sometimes still does." He is ubiquitous in the small town that is Itacaré, but this tenacity serves him well.

When I first met Irado nearly twenty years ago, he was working in a restaurant, deftly manipulating his inability to maintain personal boundaries by clowning for tourists like us, who will laugh, tip well, and come back to the familiar and welcoming time after time. To this day, I have pictures of him feeding cake to Ella, winning her over with sugary treats and crazy antics that only a two-year-old would truly appreciate. Irado is clever like this. He can smell the money connected to outsiders, but he also genuinely likes people of all walks of life. Equally important, he knows how to ride the dynamic wave of tourism here. He makes friends with the wealthiest—the French, the Swiss, the Brazilians from the south—and slowly becomes trusted by outsiders. This allows him to be in a perpetual phase of reinvention, welcoming change amid the constant that is life here. He takes care of tourists' homes when they leave the region and return, for a bit, to their real lives. He rents out paddleboards. He doggedly pursues those who will most likely never learn to surf to take lessons from him. He does what he needs to do. Now he has constructed a small house on a tract of property in the center of town that is owned by one of the region's oldest, most well-known families, a family known infamously for its colorful history and famously for its past leadership in the local environmental movement. Here, Irado lives, making his house, and his life, on someone else's property surrounded by his beloved family of dogs.

Prospecting versus Protecting

Jeff and I take a morning run, as we like to do. The streets are nearly deserted at this hour, the sun is still low, the air as crisp as it can be in the tropics. We quietly travel over the cobblestoned streets, up and down hills, the forest on our right, the pounding of the surf to our left. A few early-morning exercisers see us and nod in solidarity. We continue on, entering the forest, running along a narrow trail, over slick rocks with tiny streams running across them. It is very, very quiet, almost magical, just the sound of our footsteps as we quietly navigate protruding roots, small streams, trails of leaf-cutter ants that cross our route. We come to a big hill that has been cleared of the forest, see the path we want, and start up to the town's airstrip, ascending the steep hill so slowly it almost seems we are walking. Sitting on his father's lap at age six, Max learned how to drive here, as did a few locals whom Jeff also patiently taught. On the deserted airstrip, with an old car, the margin for error is large. This is the hard part where I always want to stop, but the open view at the top as we turn toward the ocean is worth the pain it takes to get here. At the airstrip, we run down a long dirt path, past a few houses with dogs that bark in warning to stay away, and then take a sharp right to head back into town down the *ladeira grande*, the large hill.

The view here, too, as we turn back, is magnificent. We see the ocean in the distance, but now the town, not the forest, grows closer with every step. Suddenly, across the narrow channel I catch a glimpse of something I don't recognize. It truncates the flat horizon I have grown accustomed to. It looks like something industrial. I register it in my mind. I will ask Guiomar upon our return about this metal structure that looms just offshore and seems strange, out of place.

At the end of our route, we come upon something unexpected for the relatively early hour: people neatly lined up outside a small shack on the *orla*, the waterfront where the boats are docked. One by one, a person files in while another files out. There are young and old. No one speaks much, but everyone seems quite intent on getting inside. We pause, out of breath, and ask a kid of not more than twelve what's going on. "Oh, there's a dead body inside that we all want to see," he says, "It's pretty cool."

We walk back in silence. Jeff is processing our pre-coffee run-in with a dead body. I am thinking about not only the body but the strange sight off in the distance. As Guiomar makes breakfast, we tell her about the body. She rolls her eyes and starts the blender. I'm not sure if she is surprised, embarrassed, or hiding something. I'll ask her more later, and she will likely tell me that it is due to the rising drug violence in the town, which has created warring factions between neighborhoods. But I don't push now—sometimes there are things she and I don't want Jeff to know about this place. Instead I ask, "What the heck is that thing out in the ocean?" "Oh, they're prospecting for oil," she explains. "That platform has been there for a few weeks. There are a few of those here in the area. They move about, actually." She seems to just accept this, and turns back to the tapioca she is making.

Sure enough, the next day I see yet another platform off of one of the town's main beaches. I think of how the colonial gaze that has prevailed in Bahia for more than five centuries was first cast inward from Bahia's coastline toward the pau brasil trees that furnished wood for the European market. Once logged, this gaze shifted to the vast sugarcane plantations along the coast that also fed the European taste.[4] The vision of Bahia as a place of plentiful development opportunities became entrenched within the national culture as well, which has long advocated for increased state control of natural resources, particularly those aligned with international market forces and flexible capital. Now the opportunistic gaze is turning outward beyond the coast of Bahia. It is enticed by the promise of oil, and all eyes are looking out at the ocean where this oil is found. Then I think of Catu.

For more than ten years, Southern Bahia has been touted as an ideal site for a specific type of conservation designation, called an extractive reserve, or RESEX, a federally designated conservation area that seeks to protect natural resources and biodiversity while also safeguarding people's traditional livelihoods.[5] Here in Bahia, a small movement of folks has been advocating for a RESEX in an effort to unite the region's long-standing tradition of fishing with the need to protect its biodiversity.[6] As a former fisherman, Catu

---

4. See Schwartz (1985).

5. Extractive reserves, or RESEXs, have a long and important political history in Brazil. The idea behind a RESEX is to have an environmentally protected area where limited extractive activities (harvesting crabs in Bahia, for example, or rubber in the Amazon) are permitted by the people who live in the RESEX territory. Social-environmental activist and rubber tapper Chico Mendes, from Acre, Brazil, is one of the founders of the notion of social-environmentalism and the principle that local communities are the rightful stewards of their environment and should have the rights to manage their land (Partelow et al. 2018). Mendes fought for the idea of the RESEXs, and today they have been designated throughout Brazil and are managed by different federal agencies, including the Chico Mendes Institute for Biodiversity Conservation, the Ministry of the Environment, and the Institute of the Environment and Natural Resources (Partelow et al. 2018, 72).

6. Artisanal and small-scale fishing accounts for 70 percent of total fishing in northeast Brazil, and fisheries are the most important extractive activity in Brazil (Moura et al. 2009, 618).

and a handful of his friends are at the forefront of these efforts. They have had to work hard to instill this idea in folks, to make them see how a formalistic conservation designation, often a foreign concept to the quilombolas and nativos who live up the river, can actually benefit them.

As Catu tells it, the idea for a RESEX arose almost by accident when an anthropologist from outside the region introduced this concept to regional leaders. Buoyed by this hope, gradually over the past decade, a group of fishers and other nativos began working to publicly and politically validate Southern Bahia's ecological importance. They knew well that for people here, making a living is not the main thing—in many cases it is the *only* thing. As Dona Otília says of some of the quilombolas, "When the mangrove provides, they eat. When not, they go hungry."[7]

In time, this small group of homegrown activists built a grassroots movement. Catu and other nativos became the face of the RESEX, leading community meetings to publicize what this conservation designation means, collecting data from fishers on how their catch can be used to provide evidence of the biological significance here, and, perhaps most importantly, working with state environmental officials to envision how a protected marine area in Southern Bahia might function and benefit people.[8] "Belinda, Célia, Batista, Bárbara, we are all fighting hard for this," Catu says, making sure I am fully aware that this is not a singular battle. Even environmental NGOs that have had less of a stronghold with the fishing and native populations here join in, though Catu knows full well that this is how these things work. "I have the support of the fishers, and the NGOs get to catch 'a free ride' with me," he

---

Traditional fishing methods include, for example, shallow-water fishing instead of trolling the bottom with nets, a practice that environmentalists frown upon, as many other species die in this process. See Viatori and Medina (2019) on how artisanal fishing in Peru defies governmental policies only concerned about catch load and, instead, factors in not only economic but also social and ecological aspects of fishing.

7. The quilombolas living here often harvest crabs from the mangroves, which can only legally be done seasonally. Furthermore, when a large oil spill happens, as it did in 2005, the crab populations are often wiped out for years. Quilombolas also rely on fishing. Again, however, they are competing with a lively fishing industry in the region that meets not only local subsistence needs but also the demands of the region's tourism industry. Many boats come from elsewhere and use more sophisticated techniques, so quilombolas can have stiff competition in catching fish.

8. For a study of these processes of community involvement in RESEX designation, see Weigand (2003).

says, laughing, aware that his convening power and social capital is being used as well as that there can be strength in numbers if he joins them.

At one point, it seems that the RESEX designation is close. Environmentalists, fishing associations, and academic researchers lobby at local, regional, state, and federal levels for it to pass. This is our moment, they think. A Brazilian doctoral student who works with Catu and the other *nativos* advocating for the RESEX has secured a position in the federal government, which provides hope that this cause will be remembered "in Brasília," as they say when referring to the federal government in Brazil. Without the blessing from the national government, the RESEX will not come to be. On a trip to Southern Bahia for Earth Day, the minister of the environment personally meets with RESEX advocates to learn more about the proposed marine reserve. Her interests further fuel people's belief that their case is being seriously considered. For a while, there is hope again.

But days, months, even years pass. Every time I come back to Bahia after being gone for a while, I look at Catu, and he merely shakes his head. He knows what I will ask and answers before I have to. "Not yet," he says. But he always follows with a quick, positive commentary. "But it is close. I have hope!" Then he winks and raises his fist in a sign of power.

One day I run into Catu on the streets of town. I am buying bread for our family's breakfast; he is at his usual seat at a café. "What's with those oil platforms out there?" I ask, incredulously. I know that he, unlike others here, will have an opinion on this latest development. I also know that most residents remember when a Petrobras oil spill wiped out the fish—their food—for several seasons.[9] "It's bad," Catu states. "I can't sleep without thinking about the RESEX." He sighs and lets out a low whistle. "It isn't going to be a salvation for this region or anything, but it's an important part." "But how could this happen?" I ask, my voice trailing off in despair at the oil prospecting going on here, unable to finish my sentence. Gathering myself I start again. "How could this be happening *here*, of all places?"

Catu patiently explains. Being Bahian, he is not prone to hysteria. Or perhaps he has grown accustomed to the presence of the oil-prospecting platforms. Or perhaps he is becoming resigned to the very real possibility of his

---

9. While Petrobras is viewed as a nationalist force within Brazil, it is often seen as an icon of neoliberal expansion by other Latin American countries. Today Petrobras is one of the most powerful companies in Latin America; its reach extends significantly to Ecuador, Argentina, Uruguay, and Bolivia.

failure to get the RESEX passed. Catu starts. "You see, there isn't anything here in the region, an agency, that can regulate Petrobras," he notes. "The power we have is a *municipal* power," he says with emphasis, looking intently at me to make sure I understand. "But Petrobras can go to Brasília and get a federal order, while our power is only at the municipal level. It is such a mess here. If you don't know the law, or have it on your side, you're screwed." He continues, "The RESEX is something concrete, but changing the political character of the city isn't concrete. I dream of this coming to be. I want it. I want the politics to change. And if the politics here change, I won't worry about the ocean as much because there would be channels for environmental preservation."[10]

He pauses, as if trying to work his analysis out in his own head by verbalizing to me. "But the RESEX is more than this city, the RESEX is the people, *of* the people. It can't have the face of Catu—a RESEX with my face *não presta*, it is no good. It has to have power, to be for the people for many years." He pauses again, reflecting on one of the biggest challenges here. "But the hardest thing is to get it into people's heads that this territory is *theirs*. That they have the freedom to liberate the community to be their own."

Sponsoring Fun and Forgetting

Then come the *festas juninhas*, the street parties that happen every June to celebrate prominent saints in the region, Saint John, Saint Paul. The parties make people forget life for a while.

---

10. When Catu says their power is municipal, he is referring to the municipal and local support for a RESEX that he perceives as being overshadowed by Petrobras's federal and market connections. Trying to get formal RESEX status here is an attempt to bypass this municipal level and go straight to the federal level, as its granting agency, Institute of the Environment and Natural Resources, is at the federal level. The state is not purely an apparatus but a conglomeration of processes, as Catu alludes to here. In describing the state, Trouillot notes, "It is not necessarily bound by any institution, nor can any institution fully encapsulate it. At that level its materiality resides much less in institutions than in the reworking of processes and relations of power so as to create new spaces for the deployment of power" (2001, 127). Establishing a RESEX usually takes years and extensive political, social, and environmental mobilization. In the case of a marine RESEX in Southern Bahia, for example, some have aptly described this process as a "practice of 'assemblage' and a means of reconfiguring territory, livelihoods, natural resources, identities, and labor" (Lavoie and Brannstrom 2019, 120). But only having local connections, as Catu is referring to, puts his efforts to actually establish a RESEX at a distinct disadvantage.

I am thrilled to have my best friend, Molly, coming to Bahia. We have scheduled our visit to have her arrive in the peak of the June festas. For a few weeks, people will build massive square structures of wood, fire pits that will burn each night in front of their homes. They will wear coats, jeans, wool hats even in tropical Bahia. They will dress like "country people," a strange combination of square dance attire with braids in the women's hair, exaggerated freckles drawn on people's faces, suspenders, cowboy boots. There will be special liquors from regional fruits, *jenipapo*, *cajú*, to drink with roasted peanuts and popcorn. And there will be forró bands every night, playing the regional music of Brazil's northeast and conjuring hours upon hours of dancing. As Molly loves to dance, I know this will be a good time.

Guiomar is always especially happy around this holiday period. Walking through town, she proudly points out the fire pits that line the streets. We make a plan to meet for dinner, which will be followed by music and dancing. She assures us she will provide us with the proper attire for the festivities if we don't have it. "You need to dress the part, you know," she reminds us.

A few hours later, tipsy from copious shots of homemade liquor, we walk casually toward Comprido's neighborhood, Porto de Trás, in search of a forró band. The neighborhood has been transformed into a veritable carnival of color. Brightly colored triangle flags hang across the narrow street, connecting rooftop to rooftop in a zigzag pattern. People dance, sell cheap beer, fill our open cups with more liquor as we wander by, and cheerfully pose the question that marks this holiday, "*São João chegou?* Has St. John arrived here?" We join in the festivities. Molly learns forró in record time. Jeff dances with an eighty-year-old aunt of Jitilene's, whose nimble footwork keeps him alert. I dance with a lifeguard whom our family always jokes about, a guy we have informally renamed "It's a Pleasure to Save Lives," as this is his pat response every year when we see him on the streets, slap hands, and ask how he is doing. Now we all simply refer to him as "It's a Pleasure."

Suddenly, I see it. It's a Pleasure twirls me and grabs my hand, his eyes bright with excitement, the music loudly thumping, people gyrating and sweating and reveling rhythmically en masse. There, with lights upon it over his shoulder, is a brightly colored sign. It declares "Petrobras. Sponsor of the art and culture of Brazil." Years later, people still remember that São João as

Figure 19 "It's a Pleasure" (left), otherwise known as Arilso, takes Jeff under his wing at the town party to show him how things are done around here.

being one of the best parties in years. "Remember how well-funded it was?" they say. "Now *that* was a good São João."[11]

Scrambling for Good Jobs

The highway along the coast from Ilhéus to Itacaré carries tourists, who speed along in shiny rental cars. It wasn't always this way; Marcos, one of the kids' babysitters, likes to talk of how he used to throw his surfboards on top

11. Municipal mayors rely on the support of powerful companies like Petrobras, especially when the Brazilian state is sometimes absent and other times sympathetic to the power of economic and social influence. Petrobras has long been a sponsor of social-environmental work not only in Bahia but throughout Brazil. In 2013, the company launched the Petrobras Social-Environmental Program to integrate the social, environmental, and sports initiatives it supports. With the presidency of Jair Bolsonaro, however, it is anticipated that state sponsorship of the arts will be markedly reduced.

of logging trucks and hitch a ride as the trucks traveled along the coast.[12] But today the growing tourism industry has forced illegal loggers to use other routes farther from the gaze of the federal and state environmental agencies. Now, nativos are more careful if they hitch a ride with anyone. *Gente estranha*, people not from here who can't be trusted, drive many of the cars that pass. But gente estranha or not, tourism is not only a hope, perhaps it is *the* hope for the region. Today Southern Bahia draws international celebrities.[13] The Bahian coast is known as one of the most beautiful regions of Brazil. This is not lost on people who come here. Despite the garbage on the streets, the lack of sanitation and infrastructure, the spotty service industry, it is still a paradise to many with its postcard-perfect miles of beaches and palm trees and its verdant forest.

The kids and I go to the beach for a few hours of respite. Erasmo walks past us, leading a group of Europeans as he chats to them in decent English. He knows what to emphasize and what to leave unsaid, making life here sound like a tropical 24/7 party. Days are spent guiding, walking through the forest and pointing out orchids, bromeliads, monkeys, and waterfalls, or playing soccer or capoeira on the beach. Nights are spent in luaus and parties with friends around an open fire, music playing late into the night. Love is free and flowing. There are things that Erasmo keeps to himself, however. The Europeans will never learn about things like the lifeless teenage boy we encountered on our morning run. No one will ever truly know why he was killed; perhaps he was the latest victim of the region's drug war or a personal vendetta gone awry. Erasmo won't tell them how when tourism is down, desperation crops up. A good summer is one where "everyone is working"; a bad one is the contrary. And sometimes, trails, like the very one the tourists are walking along, become dangerous as they transform into places for robberies

12. In writing about the rapidly growing tourism industry in the region, Couto notes that the history of the region can be divided into two phases: before and after the development of this road (2007, 176). Since then, however, another stretch of the BA-001 highway descending south from the state capital of Salvador was inaugurated in late 2009. This road connects to the existing highway that was built in 1997, creating a relatively easy trip from Salvador to Itacaré and Ilhéus. If the Porto Sul project (described below) comes to be, this will very likely become another significant temporal marker in the region.

13. Southern Bahia is increasingly known the world over. Carla Bruni, the wife of former French president Nicolas Sarkozy, the Brazilian model Giselle Bündchen, the American actor Sean Penn, and the French actor Vincent Cassel, among dozens of others, have passed through this region on vacation or own homes here.

or assaults. But today, in the high season when people are working, Erasmo
and his friends are taking advantage of this, for tourism is one of the best
opportunities they have for crafting a livelihood here. They recently founded
their own tourism agency called Guias Nativos, Native Guides. He hands out
business cards prolifically that simply declare his identity: "Erasmo—Guide
and Friend."[14]

Irado, on the other hand, is a lone agent amid the region's bustling tourism business. This is partly because that's how he rolls and partly because
of his overly energetic spirit that must be hard for others to corral. Perhaps
these realities are related. Irado doesn't speak other languages, and he doesn't
have the capoeira card to play when he needs to, but he does have one thing
going for him—unadulterated tenacity and grit. Many times, this works to
his advantage.

We sit on our back porch each morning trying to plan out our day. The
kids are often still waking up, and Jeff and I are often in a pre-coffee daze
when it begins. Every. Day. "Oi Papai," Irado calls from the street in a voice
that is like a high-pitched cartoon character and comes out as "Ey-pop-píe,"
holding the last syllable until it trails off. Then it begins again. "Oi Papai," with
increasing urgency. He waits. His words hit the walls of the homes around us,
lingering in the air. In addition to the screeching call, he claps his hands to
let us know he's there. Guiomar rolls her eyes. I worry the neighbors will be
disturbed by his early-morning appeals. Sound travels in these narrow, quiet
streets, and no one is up at this hour, given that some of them rolled home

---

14. When asked why he decided to become a tourist guide, Erasmo explains, "I *chose* this
profession, and few people have the freedom to choose a profession that will sustain them for
their entire lives. I've already washed bathrooms. I've washed many plates. I used to own two
pairs of pants." Despite his enjoyment of guiding and his belief in the future it will provide him,
this work is sometimes circumscribed by the difficulties of interclass relations. He asserts the
most difficult part of this job is "educating those who are educated," teaching people not only
about the natural environment but also how to be more polite to the nativos they encounter in
their tourism experience in the region.

See Couto (2011) for a discussion of how residents of Porto de Trás in Itacaré (where
Jitilene and Comprido live) have been able to capitalize on their native characteristics in articulation with the tourism industry in the region, as Erasmo is also doing. Despite the many
challenges they face, people from Southern Bahia are working to carve a niche for themselves
not just as Brazilians, or even Bahians, but also as nativos to Southern Bahia. These nativo
representations center rather than decenter their identities; in other words, nativo presence is
paramount to representing place and to producing "the real Bahia" (see Collins 2008).

merely a few hours earlier. But Irado is relentless and calls continuously until one of us acknowledges his presence and invites him in.

He is calling to Jeff, who is simply Papai, Father, to Irado. He knows that Jeff cannot speak much Portuguese and that I'm likely headed out to do my work. Jeff is the weak link, the one who might pay him to take the kids surfing or to rent his surfboard for the week he is in town. We invite him in for breakfast, to discuss the myriad possibilities of him earning money by working with us, none of which we need. He is not easily dissuaded. He takes a cup of black coffee and pours sugar into it for a good long time. He then looks up at us, eager to hear how we might make something happen. He offers to teach us what he knows, surfing or stand-up paddleboarding. And if there's something else we want to do—ride a mountain bike, take a hike through the woods, take a morning run—heck, he'll do that too. He can find us a house to rent for next time. "I have the keys to that French person's house, you know." He reminds us he is the caretaker. He is always "at our service" whether we want it or not.

While Irado scrambles after work in any way he can find it, from "watching" homes to giving surf lessons, the other dynamic here is that some of Brazil's most successful business owners from the south of the country are the biggest backers of the tourism that feeds the region. One friend of mine has a name for these guys: they are simply called "the G8," a group of eight prominent businessmen in the region who are key players in the hospitality industry as resort, restaurant, and surf business owners. "Oh, these are the guys with all of the power, with the most land around,"[15] Guiomar says, pointing them out to me in a hushed tone when they visit. Most live in São Paulo or elsewhere and come from time to time just to check in on things.

But the growing financial power exerted by investors who live far from here has not escaped the nativos' notice. They remember what it was like before paved roads, celebrities, new development opportunities, and attempted robberies on the beach or trails when times are tough. Another friend, a small-business owner and advocate for sustainable regional tourism planning, ruefully describes the rapid pace of change here. "I just wish we could close our doors to the rest of the world—just for a bit—to get our

---

15. "The G8" refers to eight men who own prominent businesses in and around Itacaré. While people generally recognize the value these wealthy Brazilians can bring to the community with their investments in the region, the geopolitical G8 label is somewhat ironic as it reveals how economic and even political capital is perceived by the people who are local to a place.

house in order." But development doesn't stop and wait for people to organize their house.

One year I return to Bahia and check in with Guiomar as usual. We run
through the litany of the latest news. Angélica, her daughter-in-law, is "doing
really well," as she characterizes it. "She's working now," Guiomar says, a simple phrase to convey that someone is serious, buckling down on the straight
and narrow path, trying their best to make a life. If you are working, things
are good. If you are not, they aren't. It is that simple sometimes. She is not
only working, but "she has a good job," says Guiomar.[16]

Angélica has always been a success case in my mind. She got pregnant by
Guiomar's middle son, Andre, and had Murilo when she was only sixteen.
Since then she has been a guerreira, raising Murilo and now his little brother,
Gui, while holding down several jobs at once. She washes clothes to make
extra money. She works in stores, in restaurants. One year, she even had a
job at a local doctor's office specializing in women's health. This time, wondering where she has landed, I inquire of Guiomar, "So where is Angélica

---

16. Tourism development and the spread of benefits from tourism is complex; "good jobs"
with elite resorts often create the appearance of local opportunities while curtailing more meaningful ways of participating in the tourist economy. While nativos are often able to work as
maids and wait staff in the restaurants of these operations, the well-paid jobs of resort administration are too commonly held by those with the English-language and business skills that most
nativos lack—people who often come from more cosmopolitan places in the South of Brazil,
like Rio de Janeiro and São Paulo. Raul, a well-spoken, affable man from São Paulo, described
how he was courted and "brought up to Bahia" to run the vacation home sales at an elite resort,
for example. Nativos' limited participation in tourism development is corroborated in a study
on the region, which concludes, "The nativos end up being excluded from the work market
because they aren't qualified, ceding this space to people from other places with a higher level
of education" (Oliveira 2007, 198). This perspective aligns with another conclusion of this study,
asserting it can bring "marked social exclusion of the local population and marginalization of
their culture. Profits from tourism are not offering a better quality of life of the poor. The policies
[surrounding tourism development] give priority to entrepreneurs, attracting more investments
without proper planning" (Oliveira 2007, 198). An environmental NGO director and surfing
school owner, often called "The German," observes how people who come to Bahia from elsewhere to engage in development here generally "want things in the form that they want them,"
meaning that people arrive with particular ideas about labor and service. He continues, "This
neoliberal form with which we are developing here is implementing a new economic model."
Yet, as we will see in the narrative to follow, despite its challenges, tourism may well be the best
hope for this region as opposed to other forms of development—the question is how to shift
the balance of power in this industry.

working, that same clinic?" She proudly informs me, "She's at the ecoresort." Though there are countless ecoresorts in the region, I know immediately that she is talking about the most famous and prestigious one, a place where the wealthy of Brazil, and the world, come to relax. Guiomar repeats, "It's a really good job." As if to further convince me, she launches into one of Angélica's crowning moments at her good job.

It was a hot afternoon, and Angélica was working in the ecoresort's kitchen. One of the rental homes called for room service, and she was on deck to make the delivery. Angélica walked across the manicured grounds, carefully carrying a tray of tropical fruits, pastries, and sandwiches. The door to her destination was already open, and as her job was to be as invisible as possible, she slipped inside quietly. There, in a chair, slept Paul McCartney of The Beatles. She stared at him. He didn't move. Quietly placing the tray on the table, she turned to face him again. Still, he didn't move. This was her moment. Slowly, slowly, she raised her flip phone to her waist and paused. He remained still, napping in his chair. Knowing this was her moment, and it would likely never come again, Angélica furtively snapped a picture of Paul, turned on her heel, and tiptoed out as silently as she had come in. "You've got to see this picture!" Guiomar declared proudly, opening her phone to find it.

Catu has pointed out to me many times that "people from the region don't insist on jobs and training." In his view, this is something they should demand when business comes knocking in Bahia—it is their "right" to truly benefit as the people who live here. Another leader in the region, a prominent figure in Itacaré named Jorge, concurs with this assessment. As we sit outside his cultural organization, we listen to drums pulsating from the stage below. Dancers gyrate in Afro-Brazilian rhythms, representing the different *orixás*, or saints in the Candomblé religion. They are practicing for their theatrical performance this evening. I know this show well. It is one of the best cultural presentations I have seen in my life. Others concur; every time my parents talk of their visit to Bahia, they speak of "that amazing cultural show with music and dance and costumes." But Jorge has an astute sense of the delicate dynamics of power and class, and as we talk about what a good job is and how one comes upon it, he laments, "It's tough here. Someone might be really good at making food, say at making pasta, and then comes along a pousada, offering regular work as a maid, and they leave their own job that they can actually do out of their house for the security of the pousada."

Even "good jobs" have drawbacks. In low season or times of tourist decline, hours are cut or inconsistent, while in high season people work incredibly long hours. I know that Angélica often works six days a week. However, other benefits balance out these challenges. The employers of expensive resorts can impart important benefits to staff that help people individually as well as collectively. When Murilo, Angélica's son, was suffering from kidney failure, for example, the owner of the chic resort where she worked had provided plane tickets to the state capital, Salvador, and then helped cover the cost of his much-needed surgery. Though Guiomar had held community-wide bazaars for months trying to raise the funds for Murilo's medical care, with a quick transfer of funds the resort owner's support surpassed the amount that had taken her months to raise, providing patronage that was critical to her grandson's health. And in ways that extend beyond personal connections, some of the region's most successful tourism operations also have the fluid capital that is essential for supporting long-standing partnerships, say with farmers who can provide organic produce, while a resort provides the flexible capital and security to grow this produce amid changes in climate or fluxes in tourism.

Keeping Some Things Quiet

I run into Catu one lazy Sunday morning when the streets are quiet after the revelry of a typical Saturday night. He relays his experience with some outsiders, with gente estranho, just a week earlier. "It was crazy," he exclaims loudly, his eyes wide. We sit at a table on the side of the main cobblestoned street in Itacaré. We are the only people up this early. As a few municipal workers sweep up piles of beer cups from the night before, Catu tells me his story, holding a piece of bread in one hand and a coffee in the other. He takes intermittent bites and sips while telling the tale that is still fresh in his memory. "Man, I was doing what I do every Sunday, playing fresco ball on the beach, when this kid comes out of the bushes and walks toward me with an outstretched gun!" he exclaims, adding, "What the fuck?" for emphasis. It works—I have never heard him swear. The scene that follows is something only Catu could navigate deftly and, most importantly, effectively. "So I lured the young kid even closer to me," he describes. As I wonder why on earth Catu would ever choose this strategy with a youth pointing a gun at him, he continues. "Then, when he was right here close to me I grabbed his wrist, broke his thumb, and promptly led him by his thumb down to the police station," he recounts, as if this interaction was no big deal for a Sunday morning. "This was a kid of only thirteen or so, from a nearby town," he explains, adding, "For God's sake! What is this place turning into?" He shakes his head, and with a smirk he concludes, "Of course, I gave him a good talking-to as we walked." "What did you say to him?" I ask. "Oh, I told him he needed to get his life in order," Catu states.

We, too, have experienced the combined effects of poverty and tourism. Fed up with the petty theft that happens regularly, we adopt a practice that is common here—we file complaints, or *dar queixa*. Sometimes we file these complaints against those we know. Sometimes we file against strangers, against those who have invaded our space in some way, like the guy on the beach with Catu. There is no other form of justice, and though we know our complaints will do nothing, they are our principled attempts to recognize that there are institutions of justice, however flawed, in Brazil. These are our feeble efforts to make things right in a place where they often go wrong. But sometimes complaints aren't easily resolved with a broken thumb and a march to the police or with a registry of something that has happened despite knowing it will do no good in the end. Sometimes these efforts lead us down other unanticipated pathways to places we wish we hadn't gone.

Max discovers the theft, which took place in our house on a Sunday afternoon. "Mom, the Game Boy is gone," he yells. Upon our return from a day out with the family, Max usually scrambles to do an inventory of all things of value in the house. He cares less about the items themselves than the fact that they are still there, that his space has been kept safe. But this day, some electronic toys and a few iPods are gone. Though not insignificant, this is the extent of the damage. My computer, with all of my work, is still in my locked bedroom, hidden under the mattress. Nevertheless, the whole family is visibly dismayed. After months and months of living here, we like to consider ourselves quasi-locals. The robbery today dispels our illusions and quickly knocks us back into the category of tourists, of people who can be exploited because they will likely never come back.

The robbery seems to have been an inside job—someone who knew we were leaving town soon for a good chunk of time and would be out of our house all day on this particular day, as well as someone who knew we had small electronics, which everyone wants here. But as we are living in a rented house inside a pousada, with loads of people passing by all the time, the suspects are ample. It turns into a family mystery as we try to figure out how someone got inside. We find a small open window in the bathroom about six feet up from the floor. An inspection of the narrow, hidden outdoor hall leading to the open window reveals small hand- and footprints on a dusty wall. "Must have been kids," Jeff assesses. Ella's eyes grow wide and tear up at the possibility of hungry children in our house looking for Game Boys to sell. Poverty here is viscerally real.

While I have let other things go, little things that go missing from time to time, now it seems, at least in principle, that we should do something. There are systems here, and they are meant to be followed. Plus, the kids are watching, and I want to show them that institutions like law enforcement are meant to right wrongs. I loudly declare I am going to the police. Jeff says my effort will be futile. "Don't waste your time," he advises, adding, "They are worse than whoever did this." Ignoring him, I get in the car and drive alone to the police station.

I make my way through the narrow streets in the early evening. Music blares from speakers in trunks of cars as kids hang out down by the orla watching other kids play volleyball. People walk home from work. Telenovelas shine through narrow doorframes open to the street. I navigate through the winding streets, arrive at the police station, and park. A bored-looking officer escorts me in, through one room into another. The first room, with haggard yellow walls, is completely empty of furniture. A medium-sized Afro-Brazilian man sits on the ground in the corner, handcuffed. He looks up at me, and I can't determine if his expression is defiant or desperate. In a doorway looms a police officer, chatting with someone hidden to me, keeping an eye on the guy in the corner. I nod and try to act nonchalant, as if walking past a man handcuffed on the floor in the corner is a normal thing.

I sit in the next room with my back to the man in the adjoining space. "OK, tell me exactly what happened," the officer starts. I begin, count by count, with the events of the theft as he writes the details by hand in a large official ledger that looks like it is from the 1950s. The door behind me quietly shuts. I continue, when suddenly my account is truncated by a few loud thumps. We both pause. The officer stops listening and writing, and I stop talking. We both seem unsure of what to do. He rolls his eyes. There is another thump, then silence. I continue, not knowing what just happened and too intent on trying to tell my story in a language that is not my own to think about much else. I am carefully laying out all of the details, when it happens again. More thumps, then silence. Again, I pause, increasingly uncertain of whether I should ignore this and continue talking or stop. By now the officer has chosen the former. He writes, I talk, and we both ignore what we are hearing.

And then, just as I start in again with some other detail I feel he must know in his attempts to locate our stolen electronics, I hear a single word. *Respira*, breathe and something clicks—the dirty police station, my silly attempts to record our family's paltry losses with a tired police officer who could not care less, the strange noises, followed by silences. I am witnessing torture a few feet from me on the other side of the wall. I speed up my story markedly, omitting details so I can leave. I am panicked by the thumps, the silences, the respira commands, and, most of all, the knowledge of what is going on. Finally, we are done. I stand, open the door, and walk past what has become a shell of a man. I try to avert my eyes yet am also drawn to look. His eyes are dark, his face bloody. His chest heaves as he gasps for breath. Shaking, I walk out of the station into the dark of the night.

I return home in hysterics and do the only thing I know to do: I recount the story to Guiomar and Jeff. Guiomar is quiet. She will not go up against the police. While we will leave soon, she lives here. She knows when to stay quiet and when to offer her commentary on life here. But Jeff is now resolute in what needs to be done. "We need to go back to the station immediately and rescind the complaint," he said. I look at him in shock, not understanding. He glares back at me and clarifies, "We don't want the police to torture anyone for a stolen Game Boy!" He is right. While the guy was present before my story and tortured for something else, I have borne witness to how some things work here.

We drive back to the station in silence. When we enter, the guy is gone. I dart my eyes to where he sat just minutes ago, trying to silently communicate this to Jeff. We walk into the next room and sit in front of the officer. He is curious as to why I have come back. "We've thought this through more," I start, and proceed to explain that perhaps we misplaced the items that we thought were stolen and he should dismiss my entire account. "We'll rescind our complaint," I tell him, thinking of all of his careful writing going to waste in the ledger on the table in front of us. In an attempt to lighten the mood, I tell him that my husband, too, works for the government. "He, too, is a sort of police officer," I say, not knowing how to exactly explain that Jeff investigates and prosecutes financial crimes. Securities and Exchange officers carry a badge, and nothing more. The officer's eyes light up with familiarity. He looks at Jeff with new respect. "Do you want to try to make sure that no one took your stuff?" he asks Jeff, hitting one hand into the other in a punching motion as if inviting Jeff to conduct his own torture session. Jeff laughs uncomfortably and assures him that, no, we are happy to withdraw our complaint. "I'm on vacation," he assures the officer.

The next day our family wanders to the beach. I am eager to jump in the ocean, to cleanse myself from the memories of the night before in a way that only the saltwater can provide. The road is familiar, a ten-minute walk from town over two small hills with forest on either side of the wide, cobblestoned street. Caboclo, one of my favorite nativo capoeiristas, comes out from his tourist stand to give us the familiar back slap and hug. "How are you?" he asks, warmly. "My mom was just asking about you yesterday!" he marvels. Our banter continues. He is making things work, he notes. I ask about the Guias Nativos tourism operation he and Erasmo set up together. "Oh, Erasmo is working on his house in another part of town now," he informs me. "Did you have a falling out?" I ask, aware of the propensity here to have ruptures that break up friendships, jobs, opportunities, marriages, and partnerships. "No, no," Caboclo assures me, "business is slow now, so he's doing his house and I'm working here, that's all." I get his family report. His girls are ten now. "I don't get to see them too often," he laments. "They live too far, and it is so expensive to get there." "There" is Europe, where his ex-wife and daughters live. It is the same with countless other nativos.

We say goodbye and continue our stroll. We pass capoerista Nino's father, who has sold fresh coconuts and *cana-de-açúcar*, sugarcane, with lime and ginger to those en route to the beach for as long as I can remember. Decades, for sure. We look at the large pile of coconuts behind him. How does he get these every morning and store them at night? I ponder. Perhaps he pays someone to sleep and watch them. At three reais for one, five for two, this is his investment he can't afford to lose. He's got the best spot in town. Now that a new ecoresort is being built nearby, he needs to hang on for a few more years before he retires. We stop to chat and make a purchase, asking, as always, about the tourism season. "May was slow here," he tells us. "Now it is getting better, though. July always picks up a bit."

We proceed toward Tiririca, a quintessential tropical beach named after the sharp grass where enslaved people used to hide as they fled their masters. I look out and see a pickup soccer game with the lifeguard, It's a Pleasure, deftly handling the ball. He seems oblivious to all who are in the water. Or perhaps he has decided it is more of a pleasure to play soccer than save lives today. A few folks practice capoeira. Women and men sunbathe while children look for water creatures in small pools along the shore. As we reach the entrance to the beach, we see a small group gossiping. A woman comes up and whispers something to Guiomar. She stops, I wait, and she looks increasingly disturbed. After a minute, we continue along. "What happened?" I ask. "Oh, they just killed the son of the guy who owns this hut," she notes, as we walk past a small purple stand selling acai bowls and chairs for the beach. A worker rushes out, offering a chair to rent, and we shake our heads no, we're fine. "What?" I ask, looking at her. The night before, the torture, the angst, is fresh in my mind. "They took him up to kilometer 6," she continues, "put a bag over his head, and suffocated him."

There is a famous saying in Brazil, "For my friends, anything, for my enemies, the law."[17] I wonder if this was the police or some other, less formal law that was being enforced with this man. Both were possible.

Balancing

There is an unusual flurry on the streets of Itacaré. People are looking for rides, carrying signs, lining up at the bus station in a way that doesn't happen on a Sunday. "What's up?" I ask an acquaintance rushing by, surfboard under one arm and sign under the other. "We're going to Engenhoca!" he exclaims. "There's gonna be a demonstration to let folks know they can't build a megaresort on the best surfing beach in town. It's fucked up, man." He shakes his head and starts down the street, then stops and calls back. "Come with us," he urges. "We need all the help we can get!"

A few months later, I read the Sunday *New York Times* travel section year in review, which features one tourism hotspot a week. There I see it. Southern Bahia is on the list of "one of the top 52 places to visit." The region has earned this sought-after spot because it is slotted as the location for Latin America's first "six-star resort" for high-end tourism featuring private

---

17. The concepts of individualism and personalism within the context of Brazilian society were first explored by Da Matta (1979). Personalism stands in contrast to individualism and is the idea that social relations are central to getting things done in Brazilian culture.

vacation homes, a hotel, restaurants, a spa. The resort first had an Indigenous name, Wanapuru, and then, for unknown reasons, suddenly one day became known as Aquapura, or pure water. I think back to the demonstration we had that day on the beach. While helicopters had passed overhead, surveying the scene and planning for the resort development in the region, people below stamped a message in the sand. It was simple and to the point; it read "SOS Bahia." They stood next to it bearing signs saying "Save Our Beaches."[18]

A leading environmentalist in the region who is contracted by the new six-star resort proudly claims that the project takes "an ecological approach to development." "It's going to be good. It's an ecoresort!" he says. I can't decide if I believe him, if this turn is good or bad for the region. "Really!" he nods his head and continues with his convincing explanation. "No large trees will be cut during construction. Over time, you won't even see the houses from the beach. After it is all built, we are even going to close the resort and have the workers spend the night in the place they built. It will be a huge party!" I wonder if his insistence on the benefits of the project is directed at me or is perhaps meant for himself. But I also know that construction jobs, and parties, can go a long way here. Tourism is a regional hope, for certain, but people are divided on what exactly this should look like and whom it should be for. While closing pristine beaches for elites is technically prohibited under Brazil's constitution, which allows for coastal access on foot, there are mixed feelings about high-end home construction on these beaches, ranging from excitement around jobs to the cries of SOS, or Save Our Beaches.

Eventually, nearly all of the vacation homes at the resort are sold to European investors. The media characterizes the resort as "James Bond–style tourism"—dramatic hills covered with lush forests that descend to one of the region's most beautiful coves with ample waves and a good break. A website publicizing Aquapura features the sounds of the Atlantic Forest while drawings of monolithic modern concrete structures fade in and out as one scrolls through the pages. Despite an intermittent building schedule, nativo guards staff the watchtowers and entry stations around the clock. When the project is under construction, large buses park at the gates to the property

---

18. Aihwa Ong (2006, 9) observes, "The complex work of NGOs everywhere [is] to identify and articulate moral problems and claims in particular milieus." As reflected in the logo of a watchful eye that represents the SOS movement here, nativos are joining with others in keeping watch over threats to Bahia's ecological and cultural landscape.

and droves of men enter and leave throughout the day. The economics of this are precarious. Laborers earn money when there is construction, but it starts and stops. At the same time, construction fuels other things, too, things that are sometimes unintended, unanticipated, or byproducts of development here.

Pablo, a leader of a local environmental NGO in Itacaré, and I sit in his office, talking about the change in the region. Even I am starting to perceive the slow gentrification of what has become my second home. "There are two types of development," he patiently explains, "*emprendimentos*, which mean ventures or undertakings, like Aquapura, and *invasões*, or land occupations. The second is often led by people who move from elsewhere to capitalize on the promise of land and jobs, but both are related." He nods and I reciprocate, communicating that I, too, am aware of the complexity, the pressure on the region's intertwined environmental, economic, and social landscape. Pablo continues, "If Aquapura doesn't provide 2,000 jobs for people once it is built, these people will then invade the city looking for work," referring to the number of people who are supposedly employed in the construction of the resort.[19] "Plus, there are other effects of this type of development. . . ." His voice trails off.

I take a ride with Guiomar along the coastal highway. Driving from Aquapura to Itacaré, we pass two bars across the street from each other. It seems strange that they are both together, as if to increase the competition, which it seems no business owner would want. Under a parking garage at one of the bars, a pool table shines under a fluorescent light, though I've never seen anyone playing pool in the hundreds of times I've passed. Across the street, a dirty flowered curtain hangs in an open window, flapping in the breeze, hiding the décor from cars that pass by rapidly. "What are these places?" I ask Guiomar. She knows I like dive bars. "Oh, those are *puta*, whore, bars. Don't even think about stopping for a beer there." She chuckles.

Approaching town, I look down at a new cropping of houses, at Bairro Novo, the New Neighborhood, that people say came to be in large part through the construction jobs linked to Aquapura. We descend from the

---

19. Rumors claim that while the resort has awarded 1,500 construction jobs, once it opens it will offer only a maximum of 400 employment opportunities. Lorecia Kaifa Roland's study of tourism in Cuba describes how corporations build tourist infrastructure in countries far from their home base, as in the case of Aquapura, which then leads to leakages from tourism development that benefit the corporation far from the place where the tourism actually is (2011, 9).

big hill at the entrance to Itacaré. Guiomar looks at the houses below on her right and cautions me, as always, "Remember, don't ever go there." She is referring to Bairro Novo's reputation as a place of violence and illegal drug trafficking. Understandably, it has some of the cheapest housing and land options in the region. As we rumble along the cobblestoned street, I look toward the neighborhood, which looks neatly constructed, organized, *not* a place to avoid. Beyond the houses in Bairro Novo, perched precariously on the hillside, is the forest. I wonder about the infrastructure. Where does the sewage go?[20] What happens when it rains on the steep dirt roads? I remember a recent conversation I had with a German guy who moved to the region a few years back. Over a beer in his bar, I had asked what his hope for this place was, this place far from Germany, a place he had chosen to call home. He had responded without hesitation: "I sure hope that the problem of growth, of disorganized expansion which has arrived in a large way here, will slow, or even stop. I think that all the rest of the stuff going on here is reversible, but disorganized growth we don't have a way of reversing."

Catu knows what those comments mean. He, too, understands the paradox of development and growth. He had once observed, "We suffer a lot here. We are growing too fast. On the one hand I hope this place keeps growing because if it falls"—meaning if development here fails—"people won't even have a pumpkin to eat." He continues, "But on the other hand, this growth isn't good. The new neighborhoods resulting from construction aren't open to other residents of the city. They are filled with opportunistic people who have just moved to the region."[21]

He was right to a degree, but not entirely. I know a few capoeiristas who earned enough money abroad to *fazer a casa*, build their house, in Bairro Novo, something they would never achieve earning tips from tourists on

---

20. A new NGO in the region, ETIV, recently noted a staggering statistic: approximately 50 percent of people in the state of Bahia and in the city of Itacaré are lacking basic sanitation, which includes running water, trash services, and sewage.

21. While these comments on how growth has ruined the area seem to contradict Catu's hopes that regional growth continues, what Catu is talking about is the ever-present need for jobs in the region. Uncovering conceptions of place from the nativo point of view demands that we pay attention to people's efforts to carve a niche for themselves in Bahia's particular social, economic, and temporal context. This also demands that we look to the seemingly trivial aspects of people's lives, the comments they make that are superficially simple but often indicative of deeper realities.

the streets of Itacaré. And just last year the mayor had decided to "open" a forested area on the outskirts of town for land occupation. At this announcement, young men, including Guiomar's boyfriend, Manoel, had gathered their things in a hurry and run to the area to stake their claim by camping on the land. I remember Manoel's eyes, wild with excitement and possibility as he grabbed clothes to camp there for a few days. "This is my chance, man," he had said, as he scrambled to catch a ride to the land. "I need to seize it." He needed to run after.

Billboards along the road taunt those who pass, as well as those who live here, that new development is coming. "This is the place for your summer home. Here you can relax on Brazil's most beautiful beaches. This is the place to spend your holidays" they promise. But over the years, Aquapura starts, then stops, then starts again. Sometimes construction is halted because the international capital fueling it dries up for a time. Other times, it stops because of environmental violations, like constructing within the federally protected limits of the coastline or ecologically important forest areas. A trail is built to traverse the coastline along secret beaches called Havaizinho, Little Hawaii, and Prainha, Little Beach.[22] This has become my favorite place to walk, to feel the dramatic convergence of forest and ocean that first drew me here. It, too, is almost a secret, a place to go to be alone. The forty-five-minute hike to the final destination, a beach called Engenhoca where Aquapura is being constructed, keeps most people away.[23]

Sergio, one of the region's well-known doctors and a longtime friend, and I sit on his porch one Sunday afternoon. We sip cachaça as we look out over an enormous tract of land he bought for a song and a prayer decades ago. He has two small springs on his property, and we have just come up from a dip in his swimming hole that is surrounded by acres and acres of primary

22. Christian Palmer writes on the connections between Southern Bahia and Hawaii, noting, "Hawaii is used not only in surf and tourist marketing as the archetypical tropical tourist destination but also becomes a source of identity for local surfers and town residents [in Itacaré]" (2017, 1).

23. Tourism also must be understood as a "mediated activity" that is "subject to a wide variety of interventions and an equally diverse array of interpretations as to the meaning of those interventions" (Chambers 1997, 3; see also Clifford 1997). Looking more deeply at how local activists in the region strategically use, reject, and recast tourism allows for openings beyond simplistic analyses of neoliberal development versus "something else" and instead allows us to examine how perceptions of place and agency are also elements of the complex tourist economy in Southern Bahia (Palmer 2014).

Atlantic Forest. "The growth here is just the beginning," Sergio muses, then laughs and ruefully admits, "Well, maybe I'll be saying this seventy years from now. You know Bahia!"

But there is another battle brewing in Southern Bahia that makes the debates about oil exploration and Petrobras-funded activities, about tourism and urban development, pale by comparison. This battle strikes at the heart of what this place is, who its people are, and what they want for their future. In fact, the region hasn't faced a battle this big since it was colonized in 1500.

Constructing "Hope and Development"

I can't remember how I first heard about it. I do know, though the years pass, it isn't going away. Rumors, facts, conjecture, opinions, dreams, fears. They come and go, flaring up, then dying down for a spell. Everyone has an opinion about it. "Are you for or against it?" "Do you really think it will come to be?" What people are talking about bears a name that sounds simple but is far from so—Porto Sul, South Port. Porto Sul is a massive infrastructure development project that aims to transform Southern Bahia into Brazil's new export hub. Yet while talk of the project permeates the region and poses a threat for some and an opportunity for others, its distinct possibility seems so far from the reality here that it is simultaneously ephemeral, elusive, and hard to imagine.

At its core, Porto Sul is planned to be a deepwater port for exporting iron ore and other mineral deposits from a mine called Caetité, which lies 500 kilometers inland in Eastern Bahia. Over time, however, the port is intended to be far more. It was originally conceived to position Southern Bahia as a prime export location for Brazil's agricultural commodity mainstays—soy, corn, cotton, products that could arrive from as far away as the Amazon via a series of cross-state railways and roadways, many of which are yet to be constructed.

So what is behind Porto Sul? How did it come to be? For one, the state of Bahia, ruled for decades by the Workers' Party, known simply as the PT, has had a long-term vision of development. And while the PT has favored a host of social and economic programs oriented at uplifting those who have long been forgotten throughout Brazil, such as farmers and lower classes, it is also known for another central tenet—infrastructure development.[24] The Bahia State website claims, "A future with more opportunities is being realized [starting to be developed] for mining in Bahia." Porto Sul publicity materials portray it as a project that is "generating hope in development for the region." As if anticipating criticism of the port's location, given the region's biodiversity hotspot status, there are claims like "Everything was thought of in terms of eco-development.... [T]his will be the largest investment in environmental conservation in the state.... Bahia will come out in the front with a new model of sustainable development." And herein lies the complexity of the Complexo de Porto Sul—the development complex of Porto Sul, as it is aptly called.

For a time, the exact location of Porto Sul is, like many of the details of the project itself, ambiguous. First it is planned for a small fishing village called Ponto da Tulha, twenty-six kilometers north of Ilhéus. Then, after more time, additional scientific studies of its projected ecological impact, and meeting upon meeting about the proposed location, the site shifts to another small fishing village twelve kilometers closer to Ilhéus, called Aritaguá. This seems to be a small victory in the larger war fought by those who are staunchly against Porto Sul, yet perhaps every victory is significant in some way.

On one of my regular trips past the new proposed location, which looks remarkably like the first one, I stop my car. A single-lane sandy road lined with a small mom-and-pop restaurant and a modest grocery store selling the basics of life here—milk, sliced deli meat, fruit, canned goods, toilet paper— leads to the ocean. I get out and walk across the narrow strip of beach to the edge of the ocean. I look to the north. There is nothing, save miles of pristine beach. Behind me to the west, hidden behind the forest, is the Lagoa Encantada, a regional APA. I look to the south and see the city of Ilhéus. Jutting out from the coast is the Port of Ilhéus, which has served to export cacao from the region since the early 1900s but is reputedly too shallow to receive the types of ships needed for exporting minerals and agricultural commodities.[25]

I look out at the pounding ocean and think about my friend Julio, a Rastafarian guy who lives in Itacaré and struggles to make ends meet by playing in a makeshift reggae band. One day as we were casually chatting on

24. See Loureiro and Saad-Filho (2019).

25. See Santos (2019) for a comprehensive study of the history of Ilhéus's port and its relation to cacao production in the region.

the streets of town, he had said, "I'm against this whole port thing, totally against it." I was surprised he even knew about the port. While the port is an ever-present topic of discussion in Ilhéus, the town of Itacaré seems far from this reality, with its surging tourism industry that at times appears oblivious to all save the height of the waves for surfing and the next party that will transform the town into a bacchanalian hotbed for a night or two. When I pressed Julio further, however, it all made sense. "My dad lives in the area where Porto Sul will be built, he has a business right on the beach selling coconuts and drinks," he said. "I grew up there." "And what does your dad think of it?" I had asked. He answered without pause, "Oh, he's totally in favor of it. He will sell his beach stand and thinks he can retire on the money he'll get." Standing on the beach, I look around, wondering which of the small beach shacks surrounding me belongs to Julio's dad. I wonder how much he will receive for his beach shack, and if this sum will see him through to the end of his life. I wonder if this new life he will forge with payment for his shack will be better or worse than what he has now.

Those who live in this region have been assured that the two elements of Porto Sul that will most affect their lives—environmental impact and jobs—are central to the project's planning and implementation. Both claims, however, are hard to assess, for now at least. The company that originally conceived of the project, Bahia Mineração, or BAMIN, has a well-laid-out publicity campaign on its website that details a variety of mitigation and social measures to demonstrate apparently good corporate citizenship: forest restoration and the creation of a legal forest reserve, environmental restoration and erosion control, air quality and noise pollution control, clean technology, archaeological restoration, and rescue of fauna, all of which will presumably be needed when construction begins. Plus, the debate about location seems resolved for now, in favor of a slightly less impactful site. Many of those who live here assert, however, that other irreversible impacts will arrive with the development project—water with mining tailings will flow from the southern state of Minas Gerais via a pipeline and empty into the area where the port is located. Fishing communities will be affected. The regional cacao companies, which also depend on this water for processing the cacao beans, may well suffer.[26]

---

26. See Santos and Santos (2021).

But no one really knows how many jobs there will be, however, *até o porto chegar*, until the port arrives. Proponents of the project claim thousands of jobs will be created during construction, while opponents say a maximum of 400 or so people will be directly employed while around 1,200 indirect jobs will be generated.[27] Tied to this employment debate is the broader question of how long the mine will remain. Mining experts say mineral extraction can happen in Caitité for fifteen to twenty years at most. Does this time frame justify a port that will be around for much longer? And what about urban expansion, like the new neighborhoods around Itacaré that rose up during the Aquapura construction project and then, when construction halted, stayed intact, making an indelible mark that has both advantages and disadvantages for the region? But there are even more profound issues underlying this project.

Shifting Capital: Bahia to Kazakhstan to China

I sit in long meetings packed with activists. It seems like they are fighting for the very future of this region. Quite possibly they are. Their activism heats up, then subsides, then rises up again as new information becomes public, like the waves of the ocean flowing in and out. A study is done on the port and the economy, or the port and the environment, and people get engaged again for a while. Or the global price of ore drops, and talk of the project subsides. Or new partnerships are born, new politicians come into office,

---

27. Yet there are discrepancies around if and how the port will bring long-term benefits to the region. Furthermore, there is speculation that jobs will largely be for unskilled laborers during the port's construction and that, once open, much of the port will be mechanized. Commenting on promotional videos on the port, Socorro notes, "Apart from the people presumably driving the trucks, there is not a single person working on the various aspects of the project. Everything is mechanized!" Ong claims that under neoliberalism, "mobile individuals who possess human capital or expertise are highly valued and can exercise citizenship-like claims in diverse locations. Meanwhile, citizens who are judged not to have such tradable competence or potential become devalued and thus vulnerable to exclusionary practices" (2006, 6–7). The citizens of Caetité have already been "devalued" as they tell of how their groundwater is contaminated with staggering levels of radiation due to mining in the region (see also Kamino, Pereira, and Carmo [2020] on the socio-environmental damage of large-scale mining in the neighboring state of Minas Gerais).

and once again, the topic of Porto Sul rises to the fore. Up and down, though never really gone. A decade passes.

The port marries political will, private sector capital, transnational interests in Brazil, and a specific vision of the future of Southern Bahia. But these interests are often hidden from view. To uncover such complex dynamics, one must mine deeply, yet still they can be elusive. While BAMIN is at the forefront of the project, there is more behind its public face. The primary investor in the port project—a company called Eurasian Resources Group—is far from here. This Kazakhstan-based mining company is sixth in the world in exporting iron ore yet is also reputedly mired in corruption charges. The iron ore and other minerals that are to be extracted from Western Bahia and transported across the state by railroad are destined for China, Brazil's largest buyer of its natural resources and commodities; the Eurasian Resources Group and BAMIN have a recent memorandum of understanding with a consortium of Chinese companies to back the port.[28] And as if the national-transnational power dynamics surrounding Porto Sul are not complex enough, the project is also affected by the intermarriage of private companies with state and federal funding for infrastructure development projects.[29] Thus, a public space, a port, may largely serve private initiatives in the end, blurring the boundaries between state and private sector roles and responsibilities.[30]

A publicity video for Porto Sul shows ships and airplanes taking off from the coast of Southern Bahia for destinations beyond Brazil, a land that was once colonized and, in a way, could well be so again. As with the long colonial trajectory of Brazil, the processing of these raw materials will largely take place overseas while the costs of resource extraction will be borne by those

---

28. Eduardo Galeano asserted long ago, "Underdevelopment isn't a stage of development, but its consequence. Latin America's underdevelopment arises from external development, and continues to feed it" ([1973] 1997, 283). This argument still applies to Bahia today.

29. While the port will be open to public use and the state government is firmly in favor of it, it will be privately managed. Political scientists Barry Ames and Margaret E. Keck observe that in Brazil, "jurisdictional confusion between state, municipal, and federal attributions is common. States, for example, construct sewage treatment plants, but municipalities often lack the money to install the sewer systems" (1998, 5). It is the ambiguity of public-private partnerships, particularly in such a pervasive manner, that is highly offensive to many people in the region.

30. Emily Yeh writes about the murky relations around "the gift of development" and the taming of landscapes as a political project between subjects and states (2013, 264).

living here. And while capitalist development is not a new phenomenon in Brazil, these present-day development strategies extend across the landscape in new ways cast through the languages of logic, progression, modernity, and opportunity. As a state-sponsored publicity piece advertising Porto Sul claims, "The Future Asks for Passage." But at the heart of this regional debate lies a competing vision of what this place is. While mining, transportation, and transnational commerce executives from Brazil, China, and Kazakhstan are lining up in favor of this project, environmentalists, family farmers, quilombolas, fishers, nativos, and even surfers are scrambling to articulate alternative development opportunities that are grounded in a particular politics of place. Their aim: to defend a different vision of development in their home.[31]

Sitting Among the Wolves

Socorro and I arrange to meet at "her office," an outdoor café in the center of Ilhéus. As always, she arrives on time, neatly dressed, organized, and ready to talk. We order two cafezinhos and launch into a lively conversation about the region. Across the street, a life-size rendition of Brazilian author Jorge Amado stares at us as we sit and chat. Amado has written tomes on the culture of Southern Bahia, touching upon cacao, labor, and the complex cultural essence of Bahia. I find myself wishing he was around now to document all that is going on here with Porto Sul.

It has been a busy time for Socorro. She is increasingly adept at not only the tactical but the philosophical aspects of activism, of motivating and mobilizing people to understand what is going on around them and, once

---

31. In using the term "alternative developments," I am drawing from various critiques of development as a way of approaching change in the world that largely flows from the Global North to the South. Alternative developments recognize the diverse perspectives of grassroots and social movement actors who are often the recipients of ideas and strategies of northern-led development schemes rather than the designers of development that makes sense to their particular historical, cultural, ecological, and economic contexts. Southern Bahia is a place where state, civil society, and private sector actors are vying for power, and these often-opposing forces show how the region is a site of contestation as well as how this reality produces new social movement dynamics among those living here. As Escobar argues, place-based struggles can be analyzed as "subaltern strategies of localization" (2001, 139).

educated, to take a stance.[32] Despite the heat of the day, Socorro patiently explains the evolution that she has not only watched but helped to craft in her tireless efforts to transform this regional debate away from a negotiation about "only" the port and rather toward a deeper, more deliberate consideration about development in Southern Bahia. "We've had various different organizing tactics, and we're learning," she reflects. I remember the first era—the "No South Port" group. It was marked by powerful photos showing dramatic protests with people wearing T-shirts marked with dead fish and Greenpeace activists adorned in skeleton costumes hanging from a bridge to show what the port would bring. "But this approach presented what we *didn't* want," Socorro rationalizes, "and we realized we had to shift the narrative to talk about what we *did* want here."

The next iteration—a group called Ilhéus Action—"helped, a bit," Socorro says. "It communicated that we need something different from Porto Sul, a unique development trajectory that rises out of what we have here." When that approach still didn't work, the group formed a new institute, called Our Ilhéus, to mobilize local citizens. "*They* need to define their vision," Socorro states, speaking of the people who live here. "*They* need to take responsibility for guiding this region. This has to be *their* fight to define what their lives look like, what types of jobs they want, what type of environment they want to live in." She pauses, then states emphatically, "I can't do this alone. I can't do that for people. But I can help bring them together. I can provoke." She winks at me. As they say in Portuguese of someone who can motivate others with their words, their vision, *ela sabe como falar*, she knows how to speak.

She continues. "And more than just this region, the Our Ilhéus Institute is connected to a broader movement called the Brazilian Network for Just, Democratic and Sustainable Cities, which spans Latin America. This

---

32. The region's exceptional history, nature, and culture were originally articulated through a network that is named not to communicate a specific issue or agenda but rather to represent a new perception of place: the Just and Sustainable Southern Bahia Coalition Network. This network was started to resist the imposition of the Porto Sul project and, through this resistance, articulate a new politics of place. The network's publicity materials note it was "born to attend to the demands of the majority of the population and citizens, from all sectors of society, in the construction of a permanent process of sustainable development that guarantees an environment oriented toward social justice, a high quality of life, responsiveness to the legitimate needs of the population, good opportunities for dignified work and professional growth, and appropriate conditions for people's free cultural expression."

helps people feel ownership over where they live, *my* city, my place," she emphasizes. "Linking this pride to broader networks is how change happens."[33] Socorro nods resolutely, and I'm reminded of how her opinion aligns with Catu's views on development here. One day, speaking of activism around the extractive reserve in Itacaré, he had said, "We are trying to show people *direitos e deveres*, their rights and their 'shoulds,' what they have the right to do and what they should do, to show they have an opinion and become involved in the politics here because otherwise the world screws them."

The Our Ilhéus Institute and the Just and Sustainable Southern Bahia Network, which Socorro is talking about, don't align with a specific issue or identity such as environmental conservation, the plight of family farmers, long-standing territorial rights, or even a prime beach for surfing that should be protected from encroaching tourism. Rather, these initiatives are meant to be mechanisms for stimulating engagement, for provoking thoughtful discussions about the local economic, cultural, and ecological characteristics and realities and thus arriving at a broader, more inclusive vision and politics of place that ideally reaches *across* Southern Bahia's diverse populations, groups, and social movements. The Just and Sustainable Southern Bahia Network also strives to promote democratic ideologies as well as mechanisms for participation. No membership fee is required, and it is open to all—individuals, civil society organizations, neighborhood associations,

---

33. One of the most powerful mechanisms for resisting Porto Sul is building on and transforming a technique that has long been used by civil society actors throughout Brazil. This technique is the *rede*, which means "hammock" or "net." While the most common translation for rede is "network," Escobar notes how translations for this concept generally fall short in conveying its richness: "The Spanish [and Portuguese] redes, more than the English term networks commonly used to translate it, conveys more powerfully the idea that life and movements are ineluctably produced in and through relations in a dynamic fashion (assemblages would be a better translation)" (2008, 25). In communicating the profound perceptions of place and life, redes typically use email and other social networking tools while also holding face-to-face meetings and periodic events connected to the interests of the redes. At the national level, redes are quite sophisticated in articulating environmental, social, and political agendas (Scherer-Warren 2006). Redes are both a concept and a political strategy that people refer to comfortably in daily conversations. Rui, for example, talks of how his organization has over time "developed redes of relations that help to realize our work."

Figure 20 Activism against the South Port Project begins with a cry for what people don't want, "No Port," and evolves to a vision of what they *do* want—to foster region-wide sustainable development.

businesses, professionals, NGOs, institutions, and foundations working toward sustainable development for the region.[34]

As Socorro and I talk, a few members of the city council who are supportive of the Porto Sul project stop to buy a coffee. They say hello to Socorro, shake her hand, and then move on to sit across from us. I look at

34. "The goal is to ... provide opportunities to all persons and institutions and give them a mechanism for reporting on their projects," according to Socorro. In keeping with this, the rede avoids purely internet-driven and class-based knowledge production. There are regular meetings anyone can attend and public demonstrations that encourage democratic participation in constructing a more inclusive vision and articulating a politics of place that transcends singular class, interest, and cultural lines. For communicating information, WhatsApp texts, phone calls, and word of mouth are used in addition to email. This network has been very active in working against Porto Sul. As Socorro tells it, "In all of the public hearings, there was significant participation by people who claimed, opined, asked questions about the project.... Students, teachers, church representatives, politicians, environmental organizations ..., and civil society organizations were united against the arrival and exploration of BAMIN and said 'no' to the attempt to use the mining of the region as a source of wealth for its shareholders."

Figure 21 Socorro (left) and colleague Ricardo in "her office"—the town plaza.

her quizzically. Before I can ask her about this polite interaction in the face of this massive regional development conflict, she remarks as if reading my mind, "Oh, I sit here in 'my office' for a reason. I sit among the wolves." She looks around and winks again. I know exactly who she is talking about: the elites of the region who are threatened by a change in the system, by new participation in the society that for so long they have dominated, by awakening. "This is the only way to make change. To be out in the open, to talk to everyone. *That's* my strategy." She nods again and gazes up at the cathedral nearby as if looking for a sign, for divine moral confirmation that this is the right way to be in the world.

A few weeks later, after I have returned to the United States, I get a message that no one wants to receive. It comes from Socorro via a voice recording over WhatsApp to hundreds of friends around Brazil and the world. This is how she is: she opens herself wide, she is transparent and factual, telling things as she sees it. "I have cancer, my friends," it begins, and continues for several minutes. She wants people to know so that they can support her as she has long supported them. It takes no more than thirty seconds after the message is delivered for offerings of help and good wishes to begin pouring in, from Southern Bahia, from around Brazil, from other countries. Now that people know, Socorro does what she does best. She picks herself up, stares cancer in the face, and begins what she knows how to do—to fight like hell with grace.

Moving across the Region

Carmo and I arrange to have lunch in a city near her property. These meetings are not easy for either of us. I usually drive for two hours to get there with at least one child in tow. Carmo does the same, but on the back of her niece Alba's motorbike. Meeting midday also means she must usually take a day off work, which I know is not an insignificant cost for a farmer. Carmo seems even more intent, more serious, today. "There's something I need to talk with you about," she had said on the phone, not giving me any more details. And so we made plans to meet.

She arrives first and secures us a table at the simple restaurant we have chosen. It is loud and busy with the noontime crowd. I worry that Ella, the child chosen for the trip today, will be bored and distracted and that we won't be able to talk. But she is with me now, and there is nothing else to do but forge ahead, so I equip her with crayons and drawing paper and hope for the best. After ordering juice and plates of chicken, rice, beans, and salad, Carmo looks intently at me and launches in with a minimum of small talk. "Aren't you looking at that port that is supposed to be developed?" she asks. I nod. "Well, I want to talk with you about it. I need to learn more." I wonder why she cares about a port that is going to be developed two hours away from her house. "What's the connection?" I start. "What are you concerned about?"

She pauses, then speaks softly. "Well, you know they are building a railroad across the state, right?" I nod. She continues, "Well, this railroad is supposed to go right through my property." She looks at me squarely. I feel tears well up inside me, knowing what this news means. She continues talking softly, too focused on communicating the stark reality of the situation to waste time crying herself or to notice my tears. "You know what they are offering me for my land?" she asks. "Five thousand reais! What am I supposed to do with 5,000 reais? I can't buy new land, I can't buy a house. I can't even *furnish* a house with this money!" she exclaims. I reach across the table and squeeze her hand. There is nothing else I can do.

I think of a conversation with the German in Itacaré a few weeks back. He had offered an interesting perspective as a foreigner, as someone who had set down deep roots in this place and, through this process, crafted an insider as well as an outsider view of the social, cultural, and economic dynamics in Southern Bahia. "This type of movement," he had said, referring to the Porto Sul project, "is an expression of the macro situation in this country. Here we are applying policies and models that were developed in other regions where the history is completely different from Brazil's. In Southern Bahia, for example, we have a transition from slavery, imperialism, from coronelismo,[35] to a democracy that came about without a fight, without a conquest, without the taste of victory in the mouths of people." He paused, then said, "So this ends up causing serious structural problems," referring to the dilemma of how the people in the region can take ownership over building a long-term, sustainable, and inclusive development trajectory that works for *them*. I remembered his parting words on that night as I stepped into the street after our talk, walking over a mangy dog lying in a gutter lined with trash, seeing a girl of not more than fourteen pushing a carriage as she tried to soothe her wailing child, smelling raw sewage that hung heavily in the night air. What the German said stuck with me: "With these structural problems, everyone suffers, and those who suffer the most are the poor."

Carmo, who lives in the interior, and Julio's father, who lives on the beach where the port will be constructed, are not the only ones whose properties will be affected by the impending port construction. People say it will touch thousands more. Rui and I travel along the road leading from his home in Ilhéus up the coast. He is in a philosophical mood, which is not uncommon. We pass the place where Porto Sul is supposed to be built. People sit in their doorways, watching the traffic go by. A group of men slap cards down on a plastic table, while nearby a group of young girls dance to a song emanating from a cell phone. The pounding of the surf just to the east off the highway seems to echo off the hills to our west. I simply can't picture this narrow potholed road supporting any more traffic than it already has. I also can't picture it being widened to accommodate the superhighway that the port

---

35. Here, the German was talking about the regional history of cacao colonels, or barons (described in chapter 1), which connects to the power of the regional elite.

will demand. But that's just it with the port. Though no one can truly imagine it, its possibility somehow still looms large.

Rui looks over at me in the passenger seat and offers his insight as if he can read my mind. "We need to have an economy in this region that is capable of sustaining its biodiversity. Who is going to sustain biodiversity? It isn't going to be the NGOs; it isn't going to be the small resources that go into nature conservation. It is going to be the reality of the economy. The regional environmental conservation goal won't be successful if it isn't balanced with an agenda of development in areas we consider strategic—ecotourism, agroforestry, transformation of cacao into chocolate, other economic income like beekeeping and forest-based economies." The unsaid challenge we don't talk about is how to both establish and sustain, as well as to scale, long-term economic options in a place where traditional development paradigms, like Porto Sul, are easier to impose than those that are new and still unimagined.[36]

As we near Serra Grande, we drive up the town's namesake, the big hill. Soon after, we turn left into Rui's property, named "The Corner of the Lord," park the car, and walk slowly toward his *mirante*, or overlook tower. We begin to climb, going higher and higher, step after step as the tower sways ever so slightly with our presence. A gentle breeze hangs in the air. Slightly out of breath at the top of the platform, we pause and look out. Far below us to the south is where Porto Sul will be built. We are silent for a good while. Rui finally remarks, continuing our conversation on what type of development is needed here and how it might actually happen, "This new agenda, we need a new manner of constructing it. An agenda that is imposed, disciplined, is

---

36. Neoliberalism is "selectively taken up in diverse contexts" (Ong 2006) that reveal different values and conceptions of how to promote sustainable development (Goldman 2005). When applied to the Global South, neoliberalism can incorporate state power, world commodity trade (including minerals), and the transformation of rural society (Connell and Dados 2014). In the case of Bahia, there is a growing hegemony of neoliberal ideologies and strategies in Southern Bahia: the presence of Petrobras, the elite Aquapura resort, and Porto Sul all indicate a balance of power blatantly tipped toward transnational development, the flexibility of capital, an emphasis on the individual, and a need for both for-profit and nonprofit sectors to subsume former state responsibilities. These cases also reveal how neoliberal development is hybrid, dynamic, and perceived differently depending on one's position. Some of the most popular new development strategies advocated across social movements in Bahia, such as ecotourism, for example, are critiqued for neoliberalizing nature (see Castree 2008; Fletcher 2010; Igoe 2017; Logan and Wekerle 2008; McCarthy 2002; West and Carrier 2004), but at the same time tourism can be nuanced, contingent, and often called for by locals as their best development option.

unsustainable over time. We know these models well.[37] They might function for some months, some years even. But they don't sustain themselves."[38] I feel simultaneously hopeless and hopeful, caught between the dilemma of what people know all too well, that they need a way out of poverty, and what they can imagine, and realistically do, to run after and make this happen.

Watching

There is a simple Portuguese phrase that is often used: *ficar de olho*, keep your eyes open.[39] Keeping watch is what you need to do if you don't trust someone. It is also what you do if you care deeply about something: you pay attention, you keep watch over it. Keeping watch is more than metaphorical.

---

37. While capitalist development is not a new phenomenon in Brazil, the opportunistic development exerted today is achieved not through past colonialisms imposed by brute force but through present-day neoliberal strategies that extend across the Brazilian landscape in new ways. These strategies are often cast through the discourse of progression, modernity, and opportunity.

38. Given the economic and political power of Porto Sul, members of the Just and Sustainable Southern Bahia Network, as well as tourism operators throughout the region, cannot help but join Socorro in being "worried and uncertain" about the region's future. However, the ideals that this *rede* is fervently working toward provide a grounded ethnographic example for geographer Noel Castree's observation that to push beyond the current structures and models of neoliberalism, which "reclaim, reconstitute, or establish capitalist class privilege and power . . . we require utopian forms of environmental praxis to help us imagine alternative possibilities, emancipatory projects, and an end to social and environmental destruction at all scales" (2007, 290–91). See also DeVore, Hirsch, and Paulson on the concept of "convivial conservation" or "efforts to establish vital interdependencies among humans and ecosystems, toward the mutual regeneration of both" (2019, 31).

39. Neoliberal ideology generates conditions for specific forms of development exerted on particular locales and the people living in these places, as is happening in Bahia today. This phenomenon also creates a space for critique, resistance, and the articulation of alternative political processes that view place quite differently. When we move beyond the question of whether neoliberalism has become a contemporary hegemony (Harvey 2007, 36), different questions emerge that may well bear more relevance for the people most affected by what are often categorized as neoliberal ideologies and strategies. In other words, while recognizing the obvious power and reach of neoliberal development strategies (Comaroff and Comaroff 2000), how does this force also give rise to redefinitions of development in ways that the people living in a place conceive of as more beneficial to their lifestyle, ecology, economy, society, and culture? Furthermore, if neoliberalism is a political project, what particular practices and political activism arise in response to this project?

Recently, the political mechanisms for vigilance have multiplied dramatically. Internet sites, blogs, chat rooms, and redes, or networks, publicize activities and produce a community of onlookers that has the ability to be increasingly attentive to and knowledgeable about threats to place. Before a meeting with a municipal secretary of education, the leader of a group of land reform settlers turns to me and whispers, "If you have your digital recorder on you, use it please. This guy needs to be recorded." In another instance, I run into Marcelo, the single administrator of the region's state park, at the airport. He seems to be in a good mood, and I ask him why. "We've been able to increase *fiscalização*, vigilance, in the park," he exclaims excitedly. "The park guards are actually catching loggers, and things are changing. We are able to keep watch."

Vigilance happens from all angles; everyone seems to be watching everyone and everything. NGOs are watching farmers for incursions into the forest; quilombolas are watching NGOs for poorly designed projects; state environmental agencies are watching illegal loggers and hunters. The growing need for vigilance, however, is also bringing these different groups together in ways like never before because of the magnitude and imminent threat and crisis presented by certain development projects. Keeping watch is what all of these groups are doing, sometimes together, sometimes apart, to monitor oil prospecting, to protect beaches that are being privatized, and to address large-scale regional development initiatives like Porto Sul.

Acting

About seven o'clock on a hot night in Itacaré, darkness settles in as people make their way home from a long day of work. People are relaxing, taking showers to wash off the day's labor, watching telenovelas, and perhaps even resting up for a late evening of parties and music that commence at midnight and last until dawn. The smells of cooking radiate from open windows. The town settles into a collective pause from the day before the activities of the night.

A small group of young men gathers in a vacant apartment; the only people older than twenty-five are Catu and I. This is a strange assortment of people at a strange time of day. One of the young guys calls the meeting to order, takes out a handwritten ledger, and reads the minutes from last week's meeting. People interject when they need clarification of a point, and Catu speaks up a number of times, guiding the group when they seem to veer off topic. They talk about distributing clothes and food to a nearby community. Some of the young men interject, "People there are hungry, for sure," while others nod in agreement. They make plans for an upcoming rock concert. The price of admission will be a bag of beans. They talk about taking a trip to Salvador to check out a recycling center they think could be good for the town.

This is a weekly meeting of what is called the Native Group of Attitude in Action (GAN'A). They wear shirts with this name, they paint murals on the town streets, they come together to watch, and when they see something that needs doing, they act. Catu had once exclaimed when we talked about activism, "We have an *obligation* to defend not only this place and our small group but all the residents of this city! We are privileged to be born here!" There are palpable tensions here, tensions that come from different conceptions of what development means and how it will shape the region. As these struggles play out, perhaps being nativo will transform from a social category to describe birthplace into an identity that is recognized for its knowledge, experience, and rootedness here. Just perhaps, being nativo will become the core of place-based political activism, of new forms of running after that will shape the future of this region.

As the meeting comes to a close, the sound of frogs in the nearby forest almost drowns out the plans the group is finalizing for the coming week.

Figure 22 Catu (far right) and friends collect food donations as part of GAN'A's work.

There will be a demonstration on one of the nearby beaches to protest it being closed to locals who want to sell food and handicrafts to tourists. One guy mentions that they should all log on to the internet to read about the latest with the Porto Sul project. "Though this is going to be constructed an hour away, it will impact our lives here if this comes to be," he states. He gives out the website address and the name of a contact for more information and warns, "Everyone who comes to this place for tourism will pass that monster of a project." The group disbands now that their jobs have been assigned, now that coming together has provided direction for action.

I walk home alone. I have kids to get to bed. Notes to write. A few hours of sleep to catch before I start it all over again tomorrow. Listening to the town come alive again now that it has had the early-evening pause, I stroll and think. Perhaps not by accident, the name GAN'A is quite similar to the Portuguese verb *ganhar*, to win. And perhaps paying attention, watching, banding together and acting collectively to address the realities here, and taking pride in place is the only way to win. As Erasmo, who is an active part of GAN'A, says to me when I admire his tattoo one day: "I've etched this place on my shoulder. It is who I am, and who I will always be."

Figure 23 Erasmo shows his pride in place through a tattoo of the region.

Creating a Politics of Place

How do local and global influences shape particular development strategies, and how are contemporary social movements and activist groups joining together to define and defend particular perceptions of place?

Southern Bahia is a site of a critical contemporary development struggle. This region is perceived as a land of opportunity, a vision that is articulated through oil extraction, elite tourism development, mining, and infrastructure development. Yet, at the same time, Southern Bahia is seen as a place of exceptional nature and culture that should arguably be off-limits to externally defined development schemes and instead be guided by carefully planned, economically feasible development approaches that are predicated on the region's prominent natural, social, and cultural features. This chapter explores how current development debates are inciting nativos (and their allies) to oppose neoliberal development projects that they perceive as threatening to their individual, as well as collective, sense of place, culture, economy, and life. It examines how nativos are engaging in ground-up, inclusive political processes and trying to craft alternative development strategies. Looking to how the politics of place take root and rise up reveals how local-level activism and social movements can be better supported and strengthened, particularly in light of powerful global development forces.

Conclusion

Walking with Social-Environmental Activists

(Re)Entering

I came to Southern Bahia to try to understand the individual stories and the collective story of this region and what it has to show us about the world at large. I continue to return. This time I am accompanied not by my children but by a group of college students and other professors. We have traveled from the Amazon Forest to the Atlantic Forest of Bahia in an effort to understand the complexities of conservation, development, and social movements in globally important landscapes. I take the familiar flight into Ilhéus, as always looking down as we approach to get a glimpse of where the ocean meets the land and the forest lies beyond. The Christ statue in the harbor is dwarfed by a new structure, a grand bridge that joins the peninsula where the airport is located with the city center of Ilhéus. I see cars backed up on the peninsula, bumper to bumper for miles on the narrow road as the plane lowers above them. I look over at my traveling companions and wonder what they see. Do they notice the large swaths of trees? Are their eyes trained to see the need for city planning to deal with the traffic? Or perhaps they are fixated on the place where the land meets the water, as many are, and fail to perceive the realities that exist beneath the surface here.[1]

We descend the stairs to get off the plane, then enter the small airport. Though the modest structure hasn't changed, it is busier than I remember yet still marked by the familiar—crowds of people holding signs for the tourist lodges that dot the coast, people eagerly looking for their loved ones to deplane. A new modern refrigerated flower vending machine stands at the exit to the baggage claim for those needing a last-minute purchase. Wealthy tourists wait at car rental counters. Their clothing, polished veneer, and care-free holiday attitude distinguish them as not from here—they are passing through as opposed to running after life in this place. We make our way out of the baggage area, parting the chaos.

I scan the crowd. As always, Guiomar is there to meet us. She has dressed up for the occasion with a new pair of jean shorts and some sassy sandals. Her hair is straightened and freshly colored, the red hennaed hues glowing under the fluorescent airport lights. She smiles eagerly, though somewhat tentatively. She is ready for our next adventure but a little nervous about what it will entail. This is new territory. After I introduce her to the students, she looks at them earnestly and says, "I'm here to guide you. I'll show you this place." I translate for them, though there are no adequate words to capture the extent to which she can do this, the depths to which they can go, if they choose. We make our way into a large van, which will take us on the hour-long drive north along the coast. We begin to enter.

Connecting

Piling into the van, I choose a seat up front by Guiomar and settle in to hear the latest on everything. We start with her family. Flaviana, her youngest,

---

1. Since 2006, I have returned to Bahia nearly every year. Most times, my kids tag along and often Jeff does too. More recently, I have begun bringing students from the University of Colorado with me. Some years we stay a good while, which allows students to engage in deeper research projects; other times I lead short courses, hoping that a week or two of rapid, intense immersion will somehow make a lasting impression. Despite the differences from year to year in age, gender, interests, and personalities, my children have come to refer to these people I bring here in the collective—they are simply "the students."

now has two small children and a new husband. She is living in Southern Brazil, helping her husband in a profession one rarely hears about in the age of online retail—he is a traveling salesman for natural medicines. "She's happy," declares Guiomar. "She has married a good man, and their kids are enormous now." I smile, thinking of the photos I get from her periodically when she visits her daughter. "It's so cold down there," Guiomar complains. "Life is different in that part of Brazil. It's very organized."

Andre, Guiomar's middle child, has returned to Itacaré for a spell to cook in one of the town's high-end restaurants. "He and his wife split up," states Guiomar, referring to his latest love. I know this also means she doesn't get to see her new granddaughter, born just a year ago. There is a slight tone of hurt underneath Guiomar's veneer of bravado. That's how she is, though: she is tough. That's how people survive here.

As the distance that time brings falls away, Guiomar switches from her family news to fill me in on the latest with the nativos and particularly the capoeiristas we know, the twenty or so young men who have followed Jamaica, their Master, for the past two decades. She stays current with these latest developments via her son Neto, who calls her from Sweden each week. This has been a year of change and challenge. Jamaica has died, leaving a gaping hole in the town's capoeira community. My last memories are of him walking proudly through the streets, his once strong and toned frame reduced to that of an old man despite his relative youth. "He never tells anyone his age," people would say, but I knew he was in his mid-fifties. Despite his obviously declining state, he had taken to wearing a peculiar, brightly colored bonnet, a smart sports jacket, and shoes. It was the same get-up night after night, his "going out" outfit of sorts. He seemed to be doing just this, "going out" as he had not done before and connecting with those he knew in a manner that was gentle and sincere. He was also more reserved than usual, which was more than a tiny bit sad. I missed his brash, uncouth demeanor, his raspy voice that unforgivingly teased all who crossed his path, his capoeira pants and confident, barefoot stride as he walked down the streets calling out and chastising all he knew like he owned the town. For decades, he pretty much did.

Guiomar fills in the news around his death. "You know, he was a stubborn old man. He had diabetes, and he didn't care for himself or let anyone else care for him, for that matter." She shakes her head. She tells me how his nightly strolls stopped and he resorted to a life alone, at home. No one saw him as he steadily declined. He stopped walking the streets. "He was a proud man," she affirms. "But all of the guys showed up to honor him after his death. People from around the world sent messages, and they had a big capoeira roda here just like the old times."

"So how's the tribe holding up?" I ask, referring to the group of men who grew up with Neto, who were, in the height of Jamaica's reign, called Tribo Unida, or United Tribe. She updates me on the people we know, person by person. Many are living in Europe but come back to Bahia from time to time. "So-and-so is married to a woman over there," she declares. "She's good to him. She supports him a lot." She continues with what "support" means: "She even bought him a refrigerator and an oven for the home he is constructing here. But we know he comes back here and has his girlfriends." Guiomar rolls her eyes. I chuckle at the oldest stereotype in the book for capoeiristas here.

She continues her stories that are pieces of facts interwoven with gossip as well as her own judgment of the situation. As the stories roll off her tongue, I am struck by the flow, the connections, the constant flux between here and there, between Bahia and beyond. In the case of many, it seems the options are to struggle to make a life here with those from elsewhere who have come to this place or, alternatively, to struggle to make a life elsewhere, with someone who has come to love Bahia and to come back here from time to time as a nativo-turned-tourist. When those who are living abroad return to Bahia, the longing, the joy they have at being home, shows forth in photos, posts, pictures of them hugging other nativos and holding their hands high in the air in front of a beach or a forest or a waterfall or a group of tourists. The boundaries between Europe and Brazil have been fluid for more than five centuries, and they are even more porous today.

"And what about Neto?" I inquire. Her son personifies this flux. "He's there for now," she says, meaning Europe, "but he's saving his money. His dream is to come back here and get a capoeira academy going. If I reopen my restaurant, he can build an academy on top. And he finally has a good girlfriend, thank God." I remember saying goodbye to Neto that day long ago, giving him a last hug. Neto has been gone for many years now, making his life out there, la fora, which involves two children by two different wives, years of struggle to earn a decent living, and a host of other issues that come with being a Bahian in a country like Sweden that is known for its order and rules. "He works for a company that helps people move. He's strong, you know," she replies, flexing her own muscle and straightening her small frame as she glances at me proudly. "And he still teaches capoeira every day. He has brought it to the people over there." We speed past the palm-lined shores with the Atlantic Ocean to the east, the students chattering or napping in the back of the van, leaving us to catch up on things like this that matter to us both. I look out and think of Neto, the best capoeirista I have ever seen, and wonder how life for him really is la fora.

Crossing Boundaries

On the first morning of the course, which brings together both professors and students from Colorado and Brazil, we gather in the open-air structure of Floresta Viva. Rui sits at the center, students surrounding him on either side. Many wear headsets; a young woman from the local university, UESC, quietly talks into a microphone, conducting simultaneous translation into English as Rui speaks Portuguese. To the right is a large nursery of seedlings neatly arranged in orderly rows, shielded from the hot sun. As we chat, farmers stop by to pick up their plants and carry them back to their properties. They are part of a project that reforests regional farms with native plants that were once abundant throughout the Atlantic Forest. I look over and wipe the sweat from my face, then turn my attention back to Rui, who is giving a lecture on the ecology of the region to the bright-eyed youth who are eager to connect across geographical and cultural borders. He closes with an invitation to the students as he talks about conservation and development in Bahia. "This is work we can't do alone," he says. "It is important that the entire society helps us think of this . . . so that we can construct things together."[2]

After the lecture at Floresta Viva, we load onto the bus and travel to the famous Caititu Reserve, where Southern Bahia's biodiversity record was registered decades ago. This is a place I have passed hundreds of times over the years but have never had permission to enter. "So this is it," I say, looking at Rui with just a bit of awe as we unload and walk to a trailhead. I imagine teams of scientists from the New York Botanical Garden diligently mapping and cataloging tree plots here. This marking is what put this region on the

---

2. Rui's observation aligns with Igoe's call to foster more "emergent encounters" (in all of their forms and complexities) that can "mix up" singular portrayals, narratives, and perceptions and bring forth alternative stories that shape our conceptions of and actions around place (2017, 112).

Figure 24 Rui explains the significance of the region to visiting students in the nursery at Floresta Viva, which grows native plants for reforestation of the Atlantic Forest.

ecological map, so to speak. As the group strolls along the reserve's trails, the trees around us bear evidence of its past. Old tags with rusty, square-headed nails expose the age of some of the markings, revealing that once, long ago, someone came here and out of this emerged a particular story about this place. Now shiny new tags dot some of the trees. Daniel, our ecologist guide and a professor at a new university that has just been established in the region, explains to the students how, long ago, biologists recorded 450 different species of trees in a single hectare, which at the time was the highest tree species diversity in the world. "This inspired decades of conservation activity in Southern Bahia," he explains. "And where does the biodiversity record stand today?" one of the students inquires. Daniel patiently explains, "Though a few other spots have since bypassed this region, the Atlantic Forest is still at the top of world records. Eighteen years later, we remeasured these plots and still found a staggering level of diversity." He motions to the trees surrounding us. "And though the reserve has changed ownership, it still maintains its protected status. It is now owned and managed by a guy we all call 'The German.'" Though this is a different German than the surfing business owner in Itacaré, I smile to myself at how, despite the growing tourism from Europe, people who come here to put down roots are still simply referred to by their country of origin. I alone know of three different "Germans." One sells honey, one runs a surfing business, and one apparently owns this reserve. There are others. *A Inglesa*, the English woman, teaches English; *o Francês*, the French man, runs a pousada. When calling people who may not remember me, I even refer to myself as *a Norte-Americana*, the North American.

I look around at the light filtering gently through the trees. The birds chirp faintly off in the distance, and the students are quiet. It is like we are in a church of sorts. Our group is accompanied by a man named Jomar Jardim, whose last name aptly means "garden." Jomar is an Indigenous man from Southern Bahia who started his career as a field assistant in this very reserve, helping to catalog the plants that made it famous. He has since earned a master's and a doctorate and now heads the herbarium at CEPLAC, the federal scientific cacao agency. As we walk along quietly, Jomar points out flowers and plants. With each discovery, he draws connections between the biology of the forest and the cultural tales that have been passed down across generations, keeping the students engaged by these connections between place and people. "One of the best *lendas*, tales," Jomar relates, "is the story of the man who has feet that are backward. He lures people into the forest, and once inside the people become entranced by the beauty, the wonder here." The students listen quietly, looking around at the vegetation that engulfs them. "But then he runs away, and when the person looks down, they can't figure out which footsteps to follow to leave the forest because his feet are turned around. Some are lost here, forever." Jomar chuckles.

We pause for some final questions from the group to Daniel and Jomar before we leave the forest. There is a swimming hole and river still to be explored, but I sense a collective reluctance to leave this hallowed ground. A student asks about incursions into the reserve. Though it is formally protected, it is also a place in which fences can be easily cut, if they even exist at all. Daniel lets out a low whistle and says wryly, "Man, they take everything here, paca, monkey, anta, capybara, and, of course, caititu," referring to the namesake animal of the reserve.[3] "They've done this forever, you know. There is no stopping them." "They" are locals who have a culture of hunting fueled by the need to feed their families in the cheapest way possible. I think about how some things have not changed along Bahia's Atlantic coast.

---

3. A researcher from the University of São Paulo recently documented rising mammal extinctions throughout the Atlantic Forest (see Bogoni et al. 2018).

Despite all that it has weathered, from colonialism forward, the biodiversity here persists. I wonder if and how long this can continue.

As we stand in the shade of the forest, a bird lands directly above us and starts to chirp. A string of ants walks below, diligently carrying leaves on their backs as they follow one after another, doing what they do. Mosquitoes try to feed on us despite our feeble efforts to wave them away. I wonder if there is a hunter observing us from afar and thinking, perhaps, that Sunday will be a better day to hunt here.

Taking Back

The next day of the weeklong course, we load the students onto a bus and travel south and inland on our journey exploring people and place. The twenty-five students and five professors descend upon a hot, crowded restaurant. We have arrived late for lunch, the owner of the restaurant hasn't been told we are coming, and he, his wife, and everyone else in the one-room family business are scrambling to create a meal for our massive group. We are collectively irritable and tired, but there is no other option than to wait, eat, and continue on with our packed agenda. I look at my fellow professors and wistfully smile. We know we have planned poorly. We have forgotten about the traffic that is growing in Ilhéus and the difficulty of a large bus navigating down narrow two-lane roads. We have also forgotten about the time it takes to get things done in a place known for a slower, more relaxed pace. Plates of lunch start to appear, and I sigh with relief.

As the meal comes to a close, I gather folks to brief them on our next stop. I have not allowed myself to think deeply about this but can no longer push it to the recesses of my mind. "Let me give you a bit of background," I start. People look at me curiously. "Remember we've talked about land reform and family farming in Bahia?" They nod. "Next we are going to visit a model land reform settlement. A place that, as they say in Portuguese, 'has everything that could go right.'" I continue, despite the noise of cars and trucks passing on the busy highway just feet from the door. "This place has strong leadership. It is well located, as people say, right along this very road." I point outside. "And for years, things did go right. The people living here created a good life through farming. They each had their own plots to work as well as communal projects that they developed together. They raised sheep. They diversified their agricultural production so they weren't only relying on a single crop like cacao. They established a fruit-drying factory. They even got contracts with the government to process and dry the fruit to serve as snacks in public schools. In essence, they built the foundation for what could have been a utopian community." I summarize years of progress into a few sentences. "But then things went awry," I explain. "People in the community started fighting. Someone poisoned the sheep. At one point, a group of people were even locked in the fruit-drying factory. Things got very chaotic and dysfunctional." The students try to follow, and I continue. "Then a few years ago, a gruesome murder took place here."

To my surprise—as well as the students'—I start to sob. I am unable to control myself as I try to succinctly relay all that has happened here, the years of history, and the deeper understanding of the dreams and hopes of this community that have gradually unraveled. One of the students puts her hand on my shoulder in an act of empathy. I continue. "Though these people are my friends, I haven't felt safe going back in years." I look at their faces as they try to grasp the significance and gravity of what I am telling them. "This visit we are about to make now is the first time I am returning." People nod, quietly shuffling their feet, seemingly unsure of what to say. "Well, let's do this!" I declare, standing and trying to lighten the somber mood I have created. We load onto the bus to go to the Frei Vantuy Agro-Ecological Land Reform Settlement just a few kilometers away.

The bus stops at the place where I have parked my car for years. The former highway police station where bored police officers would check people's documents or weigh the load capacity of trucks passing by has fallen into disrepair. But now something else has cropped up here. There are small stands where people from Frei Vantuy sell manioc, cacao, bananas, dende oil. The trail behind them, a narrow pathway that leads to Dona Lu's house, where I used to stay, is still in use. Slowly, I ascend while others follow behind me. I wonder if they are perhaps just a little bit afraid given my recent summary of all that has taken place here. Or perhaps they want to respectfully give me my space. In either case, they seem to know that this is something I need to lead.

In a few moments we are at Dona Lu's house. It is still well kept, with pots of bright red flowers on either side of the front door. The exterior has been transformed from the blue that I remember to a bright yellow, with a fresh coat of paint. I clap my hands to indicate that someone is outside. In a matter of seconds Dona Lu appears at the door. She has lost weight over the years, and her already-small frame is almost skeletal now. She wears huge dark glasses, like some version of an aged Audrey Hepburn. "Remember me?" I ask, laughing as she comes to open the door. Her face widens in a broad smile, and she comes in for a hug. We embrace as if nothing has happened, as if it was just yesterday when my daughter Maya was playing with the chicks in her backyard and Maisa was having a cup of coffee at her table as she received Dona Lu's advice on life and leadership. The years of tension between us, my memories of the murder here, fall away. I not only feel happy, I feel safe.

Slowly, we head over in the direction of the former fruit-drying factory. Underneath the five bagaças that are in a state of disrepair yet still functioning, an area has been transformed into a makeshift school. Desks are arranged in a row; cardboard posters of the alphabet and inspirational sayings line the walls. The community is ready to "receive us," as they say in Portuguese, and has set up chairs in neat rows facing a table at the front. As we file in, those who live here also start to arrive. A few young children mill about, shyly looking at the students. A curly-haired blond girl comes up to me and hands me a flower as the man beside her explains that this is Maisa's granddaughter. I hug her gently, and as I look over her shoulder, I see Maisa coming toward me. In seconds we are embracing, crying, laughing while the small crowd gathers around to watch our reunion. In time, after introductions, we settle to begin the afternoon's program. Other female leaders are here too. Mera, for one, looks older but resolute and calm as always. Maisa takes the front of the room and leans over to me, whispering quietly, "What should I say to them?" She nods in the direction of our group. "Tell it like it is," I reply. She looks at me and asks, quizzically, "All of it?" I respond, "Yes, all of it." Gradually, the tale of this community unfolds as Maisa addresses the group in her soft yet strong voice, calling upon others to complement her words as needed.

As the students are drawn in, I am aware of the deeper messages in her words, of messages that are only revealed with time and trust. I hope that the students can get a glimpse of how farmers suffer from stereotypes that position them in particular ways, that conjure, communicate, and perpetuate essentialized labels and deeply ingrained ideas about poverty, violence, anti-environmentalism, and "the other." I hope the students sense how this can limit their agency, their access to government agencies and NGOs, to those who have access to much-needed capital and projects. I also hope the students perceive the struggles these people confront as they work to mediate infighting, jealousies, and all-too-common realities of humanity, especially when living in community. And at the same time, I hope these idealistic, sensitive, and eager youth are able to witness the glimmers of hope and success as Maisa stands at the front of this room and tells her story to those who have come from far away to hear it.

Maisa starts with one of the things she is most proud of, something that she and the other women living here have fought for fervently: "One of the distinguishing factors in this community is the leadership of the women." She looks around at her friends, her colleagues, fellow survivors, and guerreiras, fighters, who have stayed to make a life here not only for themselves but for their families. The women in the room nod and smile. As the students listen intently, Maisa recounts detailed stories of how they have become leaders in myriad ways. "Now we, the women, are actually the *donas da terra*, the property owners. The men cosign our deeds with us," she explains, "but *we* are the first name on the land title. And if we get divorced, *we* get the land. It has never been like this before," she emphasizes. The settlement has grown from thirty-nine to forty-seven families. Maisa explains, "All but two of the forty-seven properties here are owned by women, and the two owned by men are because they are widowers." Several women behind her beam, as do the female students who are carefully listening to her stories and furiously scribbling notes.

After an hour, fueled by glasses of cold cacao juice poured from large plastic jugs that the community generously prepared for us, the students and I file out of the room and climb up to the top of one of the bagaças where we find cacao beans drying in the sun. The community has found some peace after years of conflict. People who chose to leave have now chosen to return. Maisa is once again the president. She explains their plans. "But we have a project," she starts, which is what people say when they mean they have a plan to make something happen, when they are running after. "Underneath the bagaças, we will reinstate our fruit-dehydrating processes. We are going to fix up the school, we hope to produce organic cacao, and we want to continue our roadside market." She points to the street beyond, where we see cars slow down, stick money out the window, and then drive away from the roadside farmers' market with regional products like dende oil, sacks of dried manioc flour, and cacao and chocolate caseira. "But we want to do more," Maisa declares. "This settlement can be a model for others. Look where it is—on the side of the road, with acres of farms as well as acres of Atlantic

Forest. Wouldn't people like to come here to spend a night to learn about the history of this place?" she asks, and many of the students nod, looking around. "We're planning for rural tourism here," she explains, "so people can come for a night, stay with us, and learn about our lives, our struggles, our triumphs." Other members of the community smile and nod in agreement.

I have heard this before, long ago, before the infighting and fleeing of the settlement by those who had the opportunity and the means to leave. Now the founders have come back and retaken this space, carrying their dreams back with them. As the students mill about snapping pictures, touching the cacao drying in the sun, looking at the impressive Atlantic Forest looming just beyond, Henny, one of the former settlement leaders, comes up to me and whispers, "It's good to be back. I only came back a year ago, you know. It wasn't safe before then." Her eyes fill with tears. "But we are starting again. We are making our life here again." I look over at Maisa, bouncing her young granddaughter on her hip. She stands tall and takes in what is going on around her, the students' processes of discovery, the community members' pride in explaining their lives to these newcomers. She is not afraid of being seen anymore, of being recognized as someone who is making her life here

Figure 25 Maisa stands on a bagaça with the Atlantic Forest in the background. Her granddaughter Yakinne touches what could be her future—organic cacao production—as students visit Frei Vantuy.

and passing along her dreams and roles—a mother and grandmother, a female leader, a farmer, an entrepreneur—in Southern Bahia.

Being in Paradise

Maisa isn't the only one relaying her story, her luta, or fight, as they say in Portuguese. Carmo, too, is still running after. Carmo, Alba, and I make plans to see each other during the field course; when the students stop to visit the nearby university, they will ride their motorbike to meet me. I wait for them patiently, standing near the university's second-floor cafeteria and gazing out at a forested area on the campus. Rain drizzles slowly, steadily. Suddenly I fix my eyes on two women walking carefully through a small treed area on campus, their heads down as they steadily approach. All of a sudden, Alba looks up. Our eyes meet, and she gives a little "woot" and throws her hands in the air. I wave excitedly from my perch.

We find a spot to gather. Alba fetches us scorching black coffees and set-tles in by Carmo's side. As the rainfall outside grows more intense, splashing onto the tables nearest to the windowless walls and slapping the pavement below us as people buzz about us ordering their lunches, I strain to hear Carmo's soft voice. I know what she has to say will be important, and I don't want to miss a word. "You ask what has changed," Carmo starts in. "Well, not much. Our biggest difficulty, as farmers here, is that we aren't united." I know she is referring to the entire region, to the various groups that are working individually rather than collectively as they strive to improve their lives. Ever the optimist, she follows this update with her dream. "But there are sixty-five small chocolate factories in the region now!" I think of how there were none when I began working here. "And I plan to become the next," she declares.[4]

Over the next hour, Carmo will speak. I will listen, furiously take notes, and provoke her with questions from time to time. Alba will quietly observe, interjecting when appropriate or when Carmo misses a detail. Each time we reunite as if we have never separated. Their wisdom runs deep, and they are generous with it.

---

4. See Altieri and Toledo (2011) on how agroecological movements in Brazil and other regions of Latin America are resisting neoliberal modernization based on large-scale agribusi-ness and promoting ground-up involvement that empowers and capacitates farmers and asso-ciations to participate in regional agricultural initiatives.

We continue with another big topic—their home. For the time being, their property is safe from Porto Sul's development footprint, which is large to be sure. "We stopped it for now, but we know they'll be back," Carmo states, talking about the threat to her land by the massive infrastructure project. Alba winks and nods in agreement, referring to the companies, the government, the myriad forces behind this project. She also clarifies an important point that I raise. Their farm was actually *not* in the pathway of the railroad for the Porto Sul project after all but within the potential boundaries of a new conservation unit that was planned as a concession for the project. This is common with infrastructure projects in Brazil—this is how the Conduru Park and the APAs came to be when the road was built long ago. Carmo continues, "They came and mapped everything, they did studies, and in the end, they made a decision that would bring a *gasto social*, a social cost, to the family farmers who would lose their land." I am relieved. "But we don't trust this. We fear they'll be back." Alba nods and says, "Pode creer," you'd better believe it. Carmo continues, "Everything begins with education, and if we don't understand that *we* are the ones who need to construct our reality, we will always be small farmers." I know, perhaps better than most, what she means by this term. I think back to when we first met, how Carmo had publicly chastised others for using the term "small farmer" and declared to the large crowd gathered on that day long ago, "This is my life, and there is nothing *small* about it!" I know she is using this term now to emphasize how people here can stay "small," can be powerless against the larger forces at play.[5]

But other things are changing, too. Southern Bahia is becoming, again, a place known for cacao and chocolate production. This is why Carmo and Alba haven't yet given up and instead continue to pursue every opportunity within their sphere of control. Carmo gives me the contact information for her neighbor, who produces organic cacao that has won prizes in Europe for the chocolate that comes from it. "He gave a three-day course, which is where

5. Later in our chat, Carmo expounds on this concept even more, which is very different from the myriad ways in which she personally pursues any and every opportunity to improve her farming processes and production. She notes that when viewed as a group, family farmers in the region don't try to band together enough to collectively surmount the obstacles that stand in their way. Instead, she observes, "they wait, for public policy, for government, state or federal, and what happens is they [farmers] lose time, and when they perceive that they could have changed, modified something to make better lives, the time has already passed."

Figure 26 Alba (left) and Maria do Carmo (right), with happiness, hope, and tenacity in their eyes.

I learned new production techniques," she tells me, and then offers sagely, "We need to *be* in a place to really understand it." She nods. "And I'm in my place." The rain has stopped, and we need to finish our time together. Alba collects their backpacks and gives me a strong hug. "Until the next time," she says with a wink. Carmo looks deeply into my eyes and leaves me with her dream. "Miracles happen twenty-four hours a day. When you come back, I want to have my small production facility going." This reminds me of when they visited long ago, carefully transporting a gift of their chocolate to me. If anyone can get there, I think to myself, *these* women can. But I hope it won't take a miracle to do so.

Months later, after our chat, I meet the chief executive officer of Dengo, one of Brazil's high-end, rapidly growing artisanal chocolate companies. I have just visited a Dengo store in São Paulo and found it impressive, to say the least—from the visuals and marketing to the product itself. The whole setup entices people to buy high-end products from cacao, from choco-late to liquors. Salespeople quietly escort customers through a variety of options, massive cacao bars with mango, coconut, and ginger added to them, all made from different percentages of cacao. The chocolate is arranged in large, roughly hewn pieces that are carefully cut, weighed, and then packaged in chic yet simple wrapping. Photos of family farmers adorn the walls, portraying stories of where these producers are from and how they farm cacao. Nearly all are from Southern Bahia.

"We have more demand than we have supply," the CEO confesses to me. "So we need to expand our work with family farmers to help them improve the quality of their cacao beans. We need to bring them along with testing and improved farming techniques." I ask if a friend of mine can attend a meeting with the farmers in the region who are already working with Dengo. I have heard that working with Dengo gives these farmers a 50 to 80 percent higher price per pound of cacao, which, given the demand for Dengo, seems a good deal for all. He graciously agrees and asks me the name of the farmers. "There are two, actually," I start. "One is Maisa, and the other is Carmo." A week later, I get a series of phone messages. "I can't go," Maisa says, the regret hanging in her voice. "I need to be in Salvador working on the permissions we need for our community's organic cacao project." But Carmo can make it. "I'll be there," she assures me. "Whatever I need to do, I'll do. And I'll share all I learn with Maisa, too, of course," she states, reminding me of her comment on how people need to unite, to band together, to have more power in production, among other things.

Carmo does just this. She spends the day at the training and subsequently shares all she has learned with Maisa. But this might not be enough to propel Carmo and Alba into a successful business as organic producers in Bahia's primary cacao region. As always seems to be the case, there are other realities to contend with. I call her a few months later and get not Carmo but Alba on the phone. "Auntie is on the farm. She's resting," says Alba. "Since you left last, her health hasn't been good. Not good at all." We chat for a while, hang up, and keep in touch via text messages. The latest message is one of the saddest I've received from Bahia in the past several years. "We may have to sell our paradise," Alba recounts. Though she fails to say why, I fear it is a combination of Carmo's health, the high toll that farm labor is taking and the "days off" that Carmo never gets, and the difficulties of breaking into the competitive organic cacao market. Alba doesn't say this all to me, though. Instead, she sends another text to clarify what she means by the term "paradise." "This is what we call our little piece of land, our farm," she says with a smile and a tree emoji.

Accompanying

The band of students and I walk along the streets of Itacaré in the direction of the São Miguel church at the center of town. We come upon Catu, running down the street with long steel poles on his back. His flip-flops slap the pavement as he passes. "Hey," he shouts at us, turning his head around to observe the group and slowing. "What are you all doing here?" he inquires with a laugh. I run up to give him a quick hand slap and hug, which he reciprocates deftly, all while holding the heavy poles. "Let's catch up soon," I say. "I'm getting my electrical training program going!" he exclaims, talking of a program he wants to start for youth in the community. "Come by later and I'll tell you all about it," he orders.

Nearing the church, I remind the students of the purpose of our visit. "This is where the first Jesuit settlers came to the region," I explain as we traipse slowly on the uneven cobblestones under the hot noonday heat and stop outside the church. "Remember what we talked about, who was already here when the priests arrived in the 1700s?" I quiz the students to see if they have been paying attention to our mobile lesson. Someone from the middle of the group offers in a tired voice, "Quilombolas and the nativos who were making their life here." I nod and say, "That's right."

Walking inside, we discover the church is undergoing a massive renovation. I smile to myself, remembering visiting here with my parents, who are devout Catholics. They had loved the church's history and simplicity and were impressed by the multitudes from the town who used this space. On this visit, my dad had remarked, gazing about at peeling paint and wooden beams in various states of disrepair, "They really need to fix it up a bit, though." This project would make him happy; restoring old churches, or building new ones, is a passionate cause of theirs. In the front of the church sits a large box with a simple sign: "Please contribute to restoring our history. Help reconstruct São Miguel." An impressive amount of coins and bills fill the thick transparent plastic box with a strong lock on top.

I had tried to find Comprido to guide us today but learned he was in the South of Brazil with his wife and child. I wonder if he has actually moved away, hoping not. Comprido is a hope for creating a dignified life based on the culture of Porto de Trás and this region. Instead, I have found another nativo, Makiba, who guides us and tells stories as we walk through the dusty, cool structure. He points out the intricate inlaid wood in the church floor.

Statues stand throughout, waiting to be unearthed from the protective plastic sheets that cover them once the structure is ready for use again. We take turns climbing to the highest point in the building, looking out to where the river and ocean meet. We hear a smooth reggae beat emanating from a stereo off in the distance, reminding us of the present, but the centuries of history in this structure are palpable as we take in the view of the forest, the river, and the ocean beyond.

As the students and I depart the dark, cool church and walk back into the bright, scorching Bahian sun, Makiba pauses and looks down at where he stands. We are hot and tired, but what he says causes us to collectively pause. "When they were building the road here," he points to the handlaid cobblestoned street below us, "they found bones buried here." The students perk up. "Lots of bones," Makiba adds for emphasis, realizing he has their attention. "We're not really sure of the story, but it seems that this used to be a graveyard of sorts." We look down at the street where we are standing in mutual confusion and fascination, imagining the bones underneath us, a situation we will likely never understand. Yet this makes it no less real. Bahia is a place that demands not only time and tenacity but also sometimes more than a little bit of faith to even begin to comprehend this landscape layered with history and meaning. "Let's move along," I urge the group, and we amble toward the modern-day urban quilombo neighborhood of Porto de Trás.

Strolling up the road, we see a woman braiding her daughter's hair. We smell noontime meals of rice and beans wafting from the many front doors that are open to the narrow cobblestone road. We walk between a row of brightly painted houses that have been constructed one after another with common walls in between. An elderly woman sits in front of her house, curiously watching our brood pass. A man delicately mends a fishing net with such concentration he doesn't look up.

At the end of the street, up a steep hill, we arrive at the Porto de Trás Cultural Center, the place where I have sat for hours with Comprido and others listening and learning to the tales of this place. The place where I have scanned the water, waiting patiently for river dolphins to appear, where my kids have taken capoeira classes. I think back to Ella, left here alone one day after her class when whoever was supposed to pick her up forgot. She was returned home by a local kid who carefully put her on the handlebars of his bike and pedaled her a few miles over the same rough cobblestone streets. This simple act of kindness was the highlight of her week. People take care of others here. In the small but dynamic town of Itacaré, Porto de Trás is a close-knit neighborhood with deep historical roots.

Today the cultural center is thriving. Kids sit on the ground, drawing publicity posters for an upcoming play that will feature the region's folklore. A young boy sweeps the hall where the performance will take place, preparing it for the crowds that will arrive. "It's gonna be packed this weekend," Makiba confirms. "We are telling one of the most famous folklore stories in the region, the Bicho Caçador, or Hunter of Beasts. You all should come," he invites us as he leads us to an adjacent room with counters and stoves. "And here's where we have our culinary school," he explains as we file along behind him. "We have famous chefs starting to come here to learn about our cuisine and to teach local folks their techniques." I've heard about this. Nativos tell of Meia Noite, Midnight, a particularly dark-skinned capoerista who has moved to São Paulo and become a well-known chef. He returns each year to hold courses for nativos, teaching them how to adapt and share their traditional recipes with the gourmet restaurants that are increasingly popping up in the town. Given the pervasive tourism industry here, this seems a viable and respectful livelihood. Perhaps the next Top Chef will be from Bahia.[6]

After our tour, the students descend from the cultural center and amble over to check out the river. I leave them to veer off in another direction. On the doorstep of the home I am seeking is a teenage girl. "Is Jitilene here?" I inquire. "She's at work," the girl responds in a bored, flat voice. "What about Dona Julia?" I continue. "Yeah, she is up there," she nods in the direction of a narrow, dark stairway. "You can go on up."

I walk into the house, gently calling out, "Dona Julia?" Ascending the stairs I immediately find her. She is an even older woman now, seated on a plastic-covered couch in front of an open window with a view to the river. She peers over at me and smiles a pleasant yet tired toothless smile. Her eyes light up, trying to place me, but I know she probably can't. I perch on the edge of the couch. Another one of her many family members sits across from us, looking curiously at this gringa who has arrived out of nowhere. After exchanging the usual pleasantries and reminding them who I am, I pepper her with questions. I learn Jitilene is working at a pousada, a small guesthouse, in

---

6. Brazilian chefs are increasingly gaining global fame, as seen in the case of Alex Atala, among others, who are famous for Amazonian cuisine. Atala has restaurants in São Paulo and New York and is one of many rising chefs in Brazil.

the tourist neighborhood. "She's doing well, but she works a lot," her mom confesses. "Eduarda is doing well, too," she replies to my questions about Jitilene's daughter. Knowing the students will be getting restless, I stand to leave. "Let me give you my latest number," I offer. "Pass it along to Jitilene, please. I'd really like to see her." I press the paper into Dona Julia's hand. Later that night, I receive a message. "Hey, how are things going, my friend?" Time and distance fall away, and Jitilene, who shyly welcomed me into her life as she was just starting it, and I pick up our relationship again.

We arrange for a visit. "I'm off on Wednesdays and one Sunday a month," she explains. The following Wednesday, she waits for me. I walk to her mother's home, and one of her many relatives directs me in, pointing her head toward a door leading to the basement. I gingerly walk down a set of five or six steep, unsteady steps and enter a tidy apartment. Jitilene is ready for me. "I've made coffee," she says, giving me a hug. I look around, impressed with her home. Off the main room lies the kitchen, with a balcony leading outside. I wonder if she can see the river during the day, but it is seven at night and dark now. On one side of the living room is a large, cavernous area that is partially dug out and open. Clearly, they plan to construct more. "Sit down," she says. I take a seat on a couch a few feet from a television playing a loud Brazilian dance program, and we begin to chat.

It seems she is well, though I wonder if she hides aspects of this truth from me. Though I had been imagining her making beds and cleaning at the pousada, I am wrong. "I'm a cook," she tells me. "I like it—it makes me happy sharing my culture, making the dishes I like. Plus, the time passes quickly." I think back to when she had proudly declared she would rather fight for her quilombo than wash dishes for someone else and am glad she is actually preparing the dishes. I also remember how people *passando fome*, going hungry, was an unspoken reality of her quilombo, Santa Amaro. It seems fitting that Jitilene has not only a good job but a job preparing food.

I look at a young girl scampering about the small room, playing with a light-skinned Barbie doll. I remember her daughter Eduarda, who is now seventeen, doing the same thing with a similar doll more than a decade ago when we met. "Whose child is this?" I ask her, assuming she is a relative's. "She's mine!" Jitilene exclaims, chuckling. "There are ten years between the girls." She chuckles again. The young one, called Maiara, is a smart, spunky pistol of a child. She dances around us, singing softly, preparing special little toasts for us to eat and bringing them out with a shy smile.

Figure 27 Jitilene in front of the river near her home.

As we converse, Jitilene's husband, Jorge, comes down the stairs. He is in from fishing. "He's a tourist guide," Jitilene says proudly, looking at Jorge. "We do alternative trips to the typical beach routes," Jorge explains. "We take people up the river, we take them fishing. We even show them the quilombo if they want." This route is indeed different from the typical tourism in Southern Bahia, which concentrates along the coastline. This type of tourism carries people inland, to the cultural sites that are less known—the waterways, the church, the communities. It travels *into* the lives of those who are making their lives here. I imagine it also reveals the region's visceral realities, how beauty and history and culture compete with poverty. "How's it going?" I ask Jorge. "Pretty well now," he notes. "Now that we got cameras put in." I look at him quizzically, not understanding. "We used to have people stealing our things, down by the orla, the waterfront. The oars we need to row, our buckets for the fish we catch. But now that we have cameras installed, we're fine," he assures me with a smile and a nod.

As I stand to leave, I ask Jitilene and Jorge if I can use their bathroom, which, of course, they agree to. There, in the humble basement apartment, is one of the nicest bathrooms I have seen in this small town. It has carefully laid tile, an enormous shower, neatly arranged bottles of shampoos and soaps. "Wow, this is pretty sweet!" I exclaim. They chuckle. "Yeah, he built me a good bathroom. He sure did." Jitilene looks sideways at Jorge and smiles; Jorge beams back at her. I feel this is a metaphor for their life together.

Observing

"Dona Otília is sick. She has a cancer," Guiomar informs me, knowing I will want to know this. "I hear she was pretty bad off and needed surgery, but I think she got it and might be getting a bit better now," Guiomar says, ever the font of information. "Well, let's visit her," I say. I never know how long Otília will be on this earth, though she is tougher than most of us despite the fact that her living situation is less than comfortable for an old woman with "a cancer," as Guiomar calls it. With my friend Magaly, a co-professor on our course, we set off for Kilometer 6, which marks the entrance to Itacaré where Dona Otília lives. From here she can survey all who come and go in this town.

As we drive to Dona Otília's house, I tell Guiomar about my plans for the coming year, of how I will do the course again. "This time you can get Comprido to help you," Guiomar states. "He's come back. He's one of the good ones," she confirms. "He has his own tourism company now and can lead you all. Erasmo is here, too," she continues. "You know he is *de confiança*, he can be counted on. He is always good with students and tourists. They say his English is pretty good now that he's lived abroad on that boat for a bit." I think back to Comprido musing how he needs to make a life here while others come, stay for a while, then move along, but Erasmo, ever the diligent worker, had left and chosen to come back. Both running after.

We pull up to Dona Otília's house, and sure enough, there she is, sitting on the porch. I hope she is happy to see us, but it is hard to tell. She is not one to show emotion. Nevertheless, she seems to have softened slightly with time and age as she allows us to hug her. I introduce Magaly and explain how she has come from the state of Acre, on Brazil's westernmost border with Bolivia—almost the farthest one could live from Bahia and still be in the country of Brazil. "How are you feeling, Dona Otília?" I ask, pointedly. She launches into a long description of her health, which culminates in just what Guiomar has reported. She was bad, she had cancer in her eye, she got

Figure 28 Erasmo shows visitors the splendor of the Atlantic Forest.

surgery, and now she is slowly on the mend. Not prone to emotion, especially directed at her personal hardships, she doesn't dwell on this.

We shift the conversation to why we are in the region. "I'm helping to teach a course on social-environmentalism in Brazil," I start to explain, looking in Magaly's direction. "She's from Acre, where we started the course," I reiterate in case she has forgotten. "Chico Mendes," Otília states loudly, naming the father of the Brazilian social-environmental movement, who hails from Acre. "I was there. I've been to Acre. I marched after Chico was killed," Dona Otília announces, referring to a famous moment in Brazilian history, in 1988, when Chico Mendes was murdered for fighting for the rights of rubber tappers in the Amazon. We quickly trace that Dona Otília knows Chico Mendes's cousin Raimundo. "Yup, we were together just after his assassination in the late 1980s," she remembers. "Next time bring Raimundo to Bahia," she orders Magaly and states, "I'd like to see him." Otília is like this. She pulls networks of associates, fellow activists, relationships, and old friends out of her pocket. I look over at Magaly, who is nodding and beaming at this incredible bond across geography and culture, from the Amazon to the Atlantic Forest, from the globally famous activist Chico Mendes to Dona Otília. Magaly recognizes we are in the presence of a guerreira. We leave and promise to return soon.

On my next visit, Otília's eye is better. We drive to her house and find her again on her front porch. She has recovered well. She looks stronger than ever and has become active in local politics again, all while monitoring Brazil's geopolitical standing. She remembers where I am from, which slightly surprises me. "There was Obama in your country," she states. "Yes, I tried to get that poem you wrote into his hands," I remind her. She continues, "Well, what's going on now with your country, with that Trump guy?" she exclaims. As I explain the dismally divided political state of the United States under Trump, she scoffs, "Agh, we've got the same situation here," referring to Brazil's current political regime under far-right President Jair Bolsonaro and launches into a diatribe about all the issues with the current president. "He oppresses women, Blacks, Indians, everyone!"

I look out at Dona Otília's property. In the garden, under a tree, is an unused stove. Someone has constructed a small structure, a house of sorts, over the stove to protect it. I think of Jitilene's cooking, of how quilombo communities were initially identified by archaeological ruins that could be dated back

Figure 29 Dona Otília on her porch.

to the era of slavery, and how food, company, stories bring people together. A low whistle from a pressure cooker somewhere inside her home and the smell of onions and garlic frying signal that lunch is being prepared. Smell is the largest trigger of memory, and I feel I am smelling a lifetime of history. I know I must clear out, for her porch is often where she eats. As I stand up to go, I notice that Dona Otília has bought a nice rocking chair for this porch where she sits, rocks, watches, thinks, and plans her next move. "I want to get another chair," she states as I compliment her on it while leaning down to hug her, "so others can be here with me."

Monitoring

Socorro, another guerreira, has emerged victorious. She has beaten cancer. Her healing was fueled by prayers, a community-wide blood drive, a period of retreat from public life and the institute she runs, family care, friend care, laughs, soup, cries, and periods of deep reflection to appreciate flowers and birds and the beauty that surrounds her in Southern Bahia! Today she stands at the front of a classroom at UESC and boldly addresses our students. "People, if you ask where else I'd rather live in the world, I'd tell you there is no other place. None at all. This is *it* for me," she affirms, nodding, cuing up a PowerPoint. The students stare back at her. They are new here, likely wondering what this larger-than-life woman is all about. "Let me tell you about this place," she begins, and starts her presentation.

Later, when we are alone, I turn to her to get the latest unfiltered story on Porto Sul. "The infrastructure project seems to be getting pushed through," she tells me with a sigh. "But we've stopped publicly talking about it. It wasn't helpful anymore. Now our job is to figure out what to do going forward." I know this is hard for her to say, to acknowledge the real possibility of a deepwater port in her backyard, her *quintal*, as they say here to express when something is very close. How do you row in one direction for a decade and then change course? When I am back in the United States, I will get a text message from Socorro: "There are a lot of things happening here, and not good things. We are going through the worst moment of the institute." I wonder what she means. Conversations like this are hard—if not impossible—to have over texts and voice messages.

One day, I see on the internet that Socorro has been interviewed about the port, as she often is. To be sure, she has backed off from her previously

Figure 30 Socorro educates visitors on the need to "stay awake" to all going on here while Rui listens by her side.

militant stance against the project due to the very real possibility that it may come to be. Ever the optimist and visionary, she talks about other forms of civic action that are rising up here. She has recently won a competitive spot as an Ashoka Fellow, joining a global group of social movement leaders. She seems more energized than ever, which says a lot.

She declares, "While years ago people started to organize to conceive of what they *did* want for this region, the activism here continues to grow stronger by the day." There are now more civil society organizations than ever, and they are paying attention to what is going on. A new initiative called Southern Bahia Global is dedicated to establishing an Agency for Regional Development, which will guide the future trajectory of the region. There's a new activist organization Sul da Bahia Viva, Southern Bahia Alive, that is finding ways to promote collective action around what it means to be alive here, as well as a group called Southern Bahia in Action.[7] And there are new businesses cropping up almost daily, a collective running after.

More than anything, there is a regional awakening of sorts, a heightened attention that if the people who live here come together, they don't need to merely accept the realities conceived of and even imposed by others but instead can collectively craft what *they* want today and tomorrow. Socorro represents this. Her palpable curiosity about the possible, sense of responsibility as a citizen, and commitment to place is predicated upon her love for it. As the port discussion continues, becoming prominent sometimes and fading into the background at others, the only thing to do is to stay vigilant. One never knows when action will be needed. Socorro closes the interview with a statement that represents her approach to life here: "We'll be paying close attention, we'll be watching, and we'll act when we need to."

Showing Up

The message comes via WhatsApp. First, it is a few photos. Rui stands with an older woman, beaming out at the camera. In the foreground is a turquoise sea with green, forested hills in the background. "Great photos! Where are you, my friend?" I ask via text. His reply comes immediately. "I'm taking my mother on a special trip," he writes. "We're in the Caribbean 'getting to know' the islands here," he says, using a Portuguese term for visiting, *conhecendo*. "The people living here, the natural environment on these islands, both are so threatened by climate change. We need to get to know these places, to understand them, before they are gone." I text back, "How did you choose the Caribbean?" Again, a quick reply. "These are the places my brother Pedro traveled before he died," he replies. "Being here connects us to him." In a strange way, this quest of Rui's also connects *me* to Pedro, the environmentalist I never met personally yet who brought me to Rui in the first place when Mario and Paula and I lay in the sailboat, looking up at the stars, the sky we share no matter where we live, just after Pedro dove too deep and never surfaced.

---

7. See Sul da Bahia Viva, n.d., "Sul da Bahia: Costa do Cacau," accessed November 14, 2021, https://suldabahiaviva.wordpress.com/; Sul da Bahia em Ação, n.d., "Sul da Bahia em Ação," accessed November 14, 2021, https://suldabahiaemacao.org.br/sul-da-bahia-em-acao/; and Instituto Nossa Ilhéus, 2022, "Principal," https://www.nossailheus.org.br/.

Our lives continue to intertwine. For many years, I have wondered if Rui could ever come to Colorado. We have long discussed this. His institute is stable. He has ample staff, two offices, and partnerships with leading universities in the United States to conduct long-term research on Southern Bahia. The space between here and there, between Colorado and Bahia, has grown thin. We are connected not only by regular visits but also by frequent text messages and phone calls. We start to plan ways to work together, to continue to bridge the gap between here and there. This possibility fuels us both.

Less than a year later, Rui and I walk down the main street of Boulder together. It is his first day here, the sun shines brightly, he and his wife, Faura, are taking in the mountain vistas. They seem happy with their choice to visit. My son, Max, now grown, has come to meet us. I sense he is curious about how these friends from Bahia are here in his hometown. We stroll together slowly, seeing the sights. On one side, a limber street performer calls for folks to watch as he squeezes his large frame into a small plastic box. Children play in a fountain that spurts water intermittently as they run and dance and scream in delight when they are unexpectedly hit with a cool stream. Parents walk with their kids, hand in hand, eating ice cream on the scorching summer afternoon.

Suddenly, Rui pauses and takes a few steps away from us. He soon comes back, carrying a letter in his hand. "What's that?" I ask, looking closer. "Look at this," he says, holding a handwritten letter, flattening it for me to read. "This was taped to a post over there"—he motions—"I just happened to notice it, and it made me curious." The envelope is marked on the front with the words "This is for you (yes, you!)." Rui opens it, and slowly I read the following message aloud:

YOU ARE THE STARS, YOU ARE THE GALAXY
It's working! Every day you're getting closer! Everything you've ever wanted is being pressed towards you. Everything is clicking—don't let illusions fool you, don't let those recent events dampen your spirits. You couldn't have more reasons to celebrate. Continue! Press on! The hardest work is done! Keep showing up, be present, open every door and let events unfold. This is your time. You can do anything.

The strangeness of this encounter leaves us speechless. It has a different meaning for each of us individually, yet at the same time it bears a collective message. Rui is thinking of his work in Southern Bahia, while Max is thinking of what he will do after college. I'm wondering how to best tell the stories that need to be told, that have been passed on to me.[8] We smile at each other and at the sheer mystery of this and continue down the road together. Rui suddenly stops and announces, "I need this energy. We have a lot of conversations, a lot of debates, back in Southern Bahia. I need to continue to find connections across difference." I smile, thinking of how one of his greatest traits is just this—listening intently, speaking his truth, and searching for common ground.

Carrying Stories Forth

Irado knows we are heading out of town soon. He also knows we'll surely be back, but this doesn't stop him from fervently tracking us down the morning we leave. There are things to be passed on, important things. Early in the morning he arrives, as usual. "Hey, Papai!" He calls, waits, and calls again more insistently. "Hey, Papai!" Guiomar rolls her eyes and goes to open the gate. I tell Max to come on down from his room. He brings his Frisbee. We all know why. We have a last meal together of coffee, tapioca, and fruit. As Irado stands to leave, he gives each family member a hand slap. He is not one for emotional hugs. "Hey," Max calls him over. "Take this for the beach," Max says, tossing Irado a Frisbee that we bought expressly for him before coming to Bahia. "Ohhhhhh, baby," says Irado as he does a little happy dance. "This is amazing!" he exclaims. "I'm gonna get me some good games going," as if we need any further assurance of this. When we return, he will be here. His hair may be bleached a different color. He may have a new dog. He will still be running after, trying to make a life through his unbridled spirit, his wit, his bigger-than-life personality.

Guiomar is quiet, as she always is when we leave. She, too, will be here when we come back. But now we have more connections in the spaces in between. We talk by text and voicemails. I look at her Facebook to see where she has been lately. We check in on birthdays and holidays. Familial bonds, as we have, are enduring.

We drive to the airport along the highway, passing all that we know. We leave the town of Itacaré along the Rio de Contas and the quilombos that lie

---

8. This reminds me of Escobar's observation that contemporary social mobilizations employ multiple strategies and emerge from multiple locations, as each of us experienced individually in this strange call to action from the universe (2017, xii).

Figure 31 Irado with all of his spunk and wit shining through.

up the river to the west. We head out of town. We pass through signs marking the state park, farms on the edge of the forest, and the two Environmental Protection Areas. We drive past the place where the port is projected to be and arrive in Ilhéus. As we cross the bridge to the airport, to my left I see the colonial city of Ilhéus abutting the coastline. To my right, I see nothing except for undisturbed mangroves and clear water. As we near the airport, the small Christ statue welcomes those who come and blesses those who live here.

As we unload our bags at the airport, Guiomar turns to me and asks, somewhat urgently, "You took your bath, right?" I know she is talking about a *banho de folhas*, a special mixture of leaves and herbs she prepares for me each time before I leave. She boils these plants in a massive pot on the stove, brings the hot mixture to me, and I drench my body with it. She knows it is my way of connecting, of arming myself with the energy of this place and bringing it where I will go. "Of course!" I assure her.

Today, in fact, I completed not only this ritual but another favorite of mine. I awoke early for a run through the forest. Ducking under vines I ran carefully, silently, as I sidestepped puddles of mud, jumped over small streams that bisect the trail, scampered past a small waterfall, and then headed up a steep hill. At the top I paused. The forest surrounded me, the ocean to my east, the town below. I heard the buzzing of insects, birds calling, a radio playing the town station from a home hidden somewhere in the forest. *This* is what the Atlantic Forest, and many forests of our world, are all about. These places, the people in them, and the movements connected to them, cannot be presented as a singular story. Landscapes and social movements alike are sites where environment, culture, economy, and even spirits collide. They are living beings. And to begin to understand another living being, you must quiet yourself and look and listen to who *they* are. You must enter, deeply.

I think about this as Guiomar passes out hugs to our family. I wipe a tear from my face, not at the leaving per se but at the pure emotion of all that this place and the people in it were, are, and will be. Guiomar looks at me and says, "Go, my friend. We'll be here when you come back, corriendo atrás." With that, she chuckles and shakes her hips in a little dance of sorts as if indicating that she knows how to not only survive but also to thrive. We hug again. Then Guiomar turns and sets off in the direction of the bus that will take her from the airport to her home, while I turn and head toward the plane that will transport me back to mine.

Figure 32 Guiomar and author before heading to the airport, friends-turned-family.

Epilogue

Fieldwork = Familywork

I have a recurring dream, something that hinges between a dream and a nightmare, actually. I am floating in the ocean at the base of steep cliffs. I can trace the rock all the way down to the sand meters below me, the water is that clear. It is a shade of blue that shouldn't be real, a light blue-green, translucent hue. Crystal-clear save a few particles that make it real, not murky, just real. I bob on the top of the water. Ella and Maya are there with me, bobbing along too. We laugh and splash and lie on our backs, looking up at the magnificent cliffs that encircle a small cove of beach.

I duck my head underwater for a second, look below, and, suddenly, I see them. A mother and a baby seal, resting on the white sand below. They are gray with white spots, which I instinctively know is strange, as only baby seals have spots. It doesn't matter. I am drawn to them. I want to, I need to, touch them. I look at the girls, take a gulp of air, and then dive down, parting the water to reach them lying there on the bottom of the ocean, quietly resting. They are still, and they let me stay with them. I reach out my hand and gently stroke their backs. They are in their moment together, I am observing and participating as I can.

And then, suddenly, I remember I have to go back up, but I cannot. I stand on the bottom of the ocean and try to propel myself upward to the air and my daughters, who swim above me, oblivious to all going on down below them. I push off the ocean floor and sink back down. I try again and am pushed back down. Suddenly, I am aware that I quite possibly lack the strength, and the air, to reach the top. The weight of the water, my nearly depleted air, I am no match for these things. I realize that my daughters will, in time, look down and notice me. I do not want this. I cannot have this. I make one last attempt, pushing off the floor with all I have, trying desperately to break through. In one last, powerful effort, I burst through that place between their reality and mine. I manage to break the surface but am frantic, panicked, gasping for air. I reach for the girls.

◆

I went to Brazil's Atlantic Forest to understand how people in this region of the world relate to each other, to understand how they conflict and converge, how cultural politics shape life in a global biodiversity hotspot. I got more than I bargained for.

As I pulled out of Boulder one December morning to head to Brazil with the family, a friend called me. We hadn't spoken for months, but she knew I

Figure 33 Setting off for fieldwork with the children in tow.

was leaving. I cheerfully told her we were in the car on the way to the airport. Her response wasn't what I had anticipated. "Be careful," she solemnly said. I naïvely brushed off this sage warning. After planning a trip that involved doctoral fieldwork, three children under the age of six, and the accompanying housing, safety, care, feeding, educating, and activity planning this brood demanded, as well as having a husband who would make the interminable trek back and forth to Brazil, I didn't think I really needed to "be careful." I had thought of most everything, and what I hadn't planned for, I'd deal with on the fly.

A few days into our transformed lives, I began to understand what she meant, but it wasn't until many months down the line—and beyond—that I really comprehended the power of my friend's sage message. As we set off with our safe and secure family unit for a different country and a new life experience, things changed. The depths of our humanity was revealed through our times together and our times apart as a family. We were challenged in ways that were not only unanticipated but also at times unwelcome, unimaginable, and even incomprehensible. At the same time, we also shared an indelible and unforgettable experience.

Post-fieldwork, as I was writing my dissertation, a visiting scholar to the University of Colorado met me for coffee. As we sat and chatted, I lamented about the difficulties of attempting to do deep ethnographic fieldwork with a family. This notion was lost on her. "I believe in 'pure' fieldwork," she said, and then explained that this meant doing anthropological research alone, unfettered, unencumbered. As if this was *ever* possible? Her comment left me reflective about the process I had experienced, which in my case was really the only way to pursue anthropological fieldwork and still have a family and home to return to.

But it went deeper than this. My fieldwork and "familywork" were inextricably linked. Often, the connections between the two were not theoretical or intellectual or even deeply psychological—they were purely practical. To conduct an interview, attend a meeting, or participate in any event for that matter, to be an anthropologist in its totality, involved a host of preparatory steps that illuminated the anthropology of my home and mothering in a foreign place. I had to make sure the children were all happy, healthy, fed, tended to, and had a plan for the day. I had to make sure there was either food in the house or money on the counter for food to be bought. If the latter, I had to decide what food needed to be purchased for the day. There were multiple days I was not my "best self," yelling as I headed out the door, "Just buy a chicken or something for God's sake—I don't care!" to Guiomar. I had to make sure the children had a ride to and from school in the event of rain (it rains a lot in the Atlantic Forest) because I was taking the car and if I wasn't in town, they would get completely soaked on the walk. In this case, they would likely get sick, and then my fragile and contingent professional anthropological endeavors would be even more compromised than they already were. I had to hire two people beyond Guiomar (remember, she informed me that she could cook and watch the baby but *not* two other kids, which made good sense given the labor of cooking in Brazil), an after-school nanny and a cleaner. We settled on a mellow Rastafarian guy to be the after-school nanny and also contracted a young woman to help with the seemingly interminable housecleaning. I then had to make sure that the nanny who would arrive after Max and Ella's school had activities planned, as the school day in Brazil lasts a mere four hours. Without a solid plan, the children would end up staring at each other for another four to eight hours until I returned home to find them watching bad cartoons on television in a language they did not yet fully comprehend. Furthermore, if this happened, as it sometimes did, I would also return to three small people and their caregivers and our household in emotional and organizational mayhem.

I had to make sure that cell phones were charged and that we all had adequate phone credit to communicate in case of an emergency. I also had to make sure that Guiomar and anyone else in the home were all getting along, which was a rare occurrence in a very small, often very hot house where everyone knew everyone's mood and business and, if they didn't, they made it their business to learn it. I had to make sure the fuscinha without seatbelts was running that day because if the car was broken, I had to send it to a repair shop where one of the people working at home (this was, in a gendered manner, usually delegated to Guiomar's boyfriend or the Rastafarian nanny) had to personally sit for the day to make sure parts weren't stolen in the process of "fixing" the car. If the car wasn't running, which was often the case, I had to stick to locally based fieldwork or spend three hours traveling by bus. One way.

Finally, I had to try to ensure that if Jeff called, the kids were around to talk with him since he was far away and feeling very cut off. It was frustrating, to say the least, for him when I stammered that I was busy and trying to get out the door and asked him if he could please call back twelve hours later. At this point, I fervently hoped it would not be raining and the phone lines would be working and the kids would be happy and amenable to talking so that we could make a human connection across time and space, both physical as well as emotional. But all of this task-tending was about life, not about my fieldwork.

Or was it? As I sat recounting this dilemma to the visiting scholar, I was overcome by the irony of it all. I went to Bahia to study conflict and difference within and among social movement and activist leaders and ended up not only contending with it in my own life but also grappling with it, crying over it, staring it in the face. Conflict and difference were present in the subject of my fieldwork and the people I studied. They were also present within me. But beyond blurring my fieldwork and my familywork, making both of these endeavors challenging and exhilarating all at once, these forces served as way more. I found that conflict and difference were *productive*. They served as rich locations, places in which to dive deeply, as I had in my dream, to grapple with, to aim to understand, and then to resurface—anew—with new appreciation for connection, understanding, survival, creativity, possibility, and hope.

Figure 34 Max, Maya, and Ella on the famous fuscinha.

Reflecting on Theory and Method

Thinking through Power: Leaders, Relations, Cultural Politics, and New Frameworks for Political Ecology

Ethnographic study challenges us to discern how people conceive of themselves and the world around them. It also calls for uncovering new ways of seeing, hearing, and thinking through the complex dynamics that emerge from paying close attention to people and processes in contested landscapes like Southern Bahia.[1] How do we come to know people in a place but also analyze *across* important local-to-global forces like social movements that are shifting, transitory, and dynamic? How do we choose analytical frameworks that are rooted within but have currency beyond a single leader and locale? How do we identify and make sense of distinctive histories, conflicts, and differences between groups with overlapping interests? How do we uncover and examine the roots of power inequities and use this knowledge and understanding to transform or construct new platforms of possibility, unity, and change?[2]

---

1. As geographers Piers Blaikie and Harold Brookfield note, for example, "while the physical reasons why land becomes degraded belong mainly in the realm of natural science, the reasons why adequate steps are not taken to counter the effects of degradation lie squarely within the realm of social science" (2015, 2). On the dynamic and contested relations between environmental conservation and development, see Peluso and Watts (2001).

2. These questions emerge from the notion of social movements as cultural entities, which first gained currency in the 1990s. Alvarez, Dagnino, and Escobar ([1998] 2018) draw from vari-

To answer such questions, we have journeyed across the dynamic and nuanced landscapes of meaning in Southern Bahia by turning to the actors and social movements in this region—environmentalists, family farmers and land reform activists, quilombolas, and nativos. These people and their social movements translate the tensions, as well as the possibilities and imaginaries, for the future of life here. They also reveal four interwoven themes.

First, power is located, tested, and contested not through amorphous social movements but rather in the "mechanisms of mediation"—the perspectives and practices of the people leading, representing, and conceptualizing these movements.[3] Turning to and drawing deeply from leaders of social movements, "organic intellectuals" who both hold power and reveal important ways of thinking about the broader social movements of which they are a part, can, in turn, be empowering.[4] In this way, social movement leaders are important conduits for, subjects to, and agents in shaping and reshaping power relations.

Second, from the vantage point of social movement leaders, we can perceive how power is relational: it arises out of interactions and negotiations.[5] Social movement leaders and their broader movements exist not in isolation but rather as intersectional and contingent cultures in and of themselves. They also act in relationship to other movements as well as with broader regional, national, and global forces. Considering this, conflict and difference reveal disjunctures that are important to pay attention to. Conflict

---

ous case studies across Latin America to reveal connections between social movements and cultural politics and Escobar (2005) points to the value of looking to the cultural politics of social movements through everyday life, historicity, politics, identity, and articulation. Salman and Assies argue that culture is "not only the shared discursive and practical backdrop for the parties in (political) conflict, it also, in its heterogeneous, globalized and contested nature, is one of the very substances of the conflict" (2017, 64). See also Davis (2012); Johnston and Klandermans (2013); McAdam (1994); Swidler (2000); and Tarrow (2011) on social movements and culture.

3. Kempf (2019, 268). In their in-depth analysis of power within the field of political ecology, Ahlborg and Nightingale assert power is relational and productive: it "generates contradictory effects within the same actions, we are able to show how resource governance processes can empower and create new relations of domination at the same time" (2018, 382). This perspective has emerged in the stories throughout this book.

4. Drawing on the work of Antonio Gramsci (2005), I am interested in the perspectives of "local organic intellectuals," the people who are advocates for and adept in practicing cultural politics that are rooted in people and place.

5. Ahlborg and Nightingale (2018, 383).

and difference also indicate potential areas of convergence. These forces exist within, alongside, and even outside of social movements when they are examined together. Social movement actors also intersect with each other, forming both loose and dense networks of activism, which have bearing on the social, cultural, economic, political, and environmental landscapes of a region.[6] And sometimes, and in some ways, the intersection of and within these forces can bring forth new relations and interconnections *across* difference, giving rise to what anthropologist Anna Lowenhaupt Tsing calls "new arrangements of culture and power,"[7] or what philosopher and political activist Cornel West deems "the new cultural politics of difference," forces that are "contingent, provisional, variable, tentative, shifting, and changing."[8]

Third, the cultural politics of social movements enable us to examine and perceive the ways in which power relations are both constraining and enabling. Looking to social movements from a cultural lens allows us to see how historical legacies as well as past and present relations of class, race, and gender shape actors and social movements.[9] These perspectives also allow us to discern their relationships with each other and with other entities like civil society, the state, and the private sector. The sometimes covert, sometimes overt cultural politics of social movements serve as places to start, but then depart from, as we try to understand broader complex and layered landscapes of meaning in dynamic environments like Southern Bahia's Atlantic Forest. Leaders, relations, and the cultural politics of social movements both individually and collectively constitute landscapes and livelihoods, rural and urban nexuses, imaginaries and possibilities for the future.

---

6. I take a "bottom-up" approach to social movement activism, drawing on individual actors and, in turn, tracing their influence upward and outward to the broader movements of which they are a part. See Porto-Gonçalves and Leff for a rich discussion of how these struggles can lead to transformation, to what they describe as nothing less than "re-signification and re-affirmation of cultural identities in their struggles for re-appropriation of nature and re-territorialization of their life-worlds" (2015, 65).

7. Tsing (2011, 245). Tsing also claims that differences can invigorate, or give life to, social mobilizations, a concept that motivates my analysis throughout this book.

8. West (1990, 94). Leff also applies the politics of difference to the Latin American context, noting that this concept goes further than merely recognizing difference. Instead, it can be a launching point for "reopening history to utopia, to the construction of differentiated and diverse sustainable societies" (2015, 50).

9. See Crenshaw (1990).

They reveal new arrangements of power *and* practice as well as new arrangements of power *in* and *through* practice.[10]

Fourth, while social movements have been studied both broadly as well as specifically, they are less often analyzed for their *intersections*. This orientation demands "thinking differently." It calls us to tease out and discern unique analytical categories that emerge from and apply to a particular context. Looking within, and across, social movements active in regions like Southern Bahia allows for focusing on the ideas, issues, and features that rise to the fore when leaders discuss and do their work.[11] For example, to understand environmentalists who influence and are influenced by broader regional, national, and global movements, analyzing specific "narratives" can help us to understand the ideologies and worldviews held by those within as well as apart from the environmental movement. To understand the realities of family farmers and land reform activists, we can perceive how subjectivities both constrain social movement actors through particular stereotypes and also provide them with a platform from which they can form new identities and reposition themselves in new ways.[12] To uncover the present-day realities of historically rooted quilombo communities and quilombola people, it is important to uncover how past and present forms of oppression constrain these individuals and their broader communities, to uncover "spaces of silence" and omissions, often within the very structures intended to support them. And to elucidate the dynamic realities of nativos in Southern Bahia, we can ascertain how neoliberal development is envisioned and exerted while at

---

10. Donald S. Moore, Anand Pandian, and Jake Kosek explore how an understanding of the politics of difference relates to identity, social exclusion, and rights within multicultural societies (2003, 49), while Clare, Habermehl, and Mason-Deese (2018) examine the power over and power to produce territories, as is evident in Southern Bahia. See also Hale (1997) and Biersack (2006).

11. Escobar's (2008) exceptional ethnography on how history, geology, biology, activists' practices, capital accumulation, development and conservation strategies, and cultural politics shape Afro-Colombian social movements was published as I was writing my doctoral dissertation, which served as the basis for this book. New, creative, and locally contextual frameworks of analysis like this are crucial for advancing social movement studies both within anthropology and across other disciplines.

12. As was explored in chapter 3, subjectivity exists within social movement cultures and reveals "differences of interests, access, power, needs, desires, and philosophical perspectives" (Biehl, Good, and Kleinman 2007, 8).

the same time gives rise to new imaginaries and forms of resistance within an emerging politics of place.

These frameworks of analysis are embedded within the rich and ever-evolving field of political ecology, which acknowledges the myriad ways in which human–environment relationships are culturally mediated and co-constructed in different contexts.[13] Political ecology pays close attention to the relationships between material and discursive struggles, to difference, place, and power, all of which emerge as themes in the preceding chapters.[14] Poststructural political ecology in particular emphasizes not only the power of cultural politics but also the ways in which people's identities, contingencies, and relations shape the cultural politics that are embedded within social movements.[15] And when applied to the Global South, political ecology can be situated in a politics of difference that can be instrumental in revealing and constructing important strategies, mechanisms, and processes for decolonizing knowledge and reappropriating territories according to the perspectives and practices of those living in these places.[16] In this way, using new frameworks for thinking through social movements on their own and in relationship with each other—frameworks that can be adapted to and taken up across historical, geographic, and cultural contexts and academic disciplines—helps us to better understand how

---

13. I argue for conceptualizing and using political ecology in creative ways, in particular through innovative storytelling methodologies, such as Goldman's (2020) use of a Maasai meeting format, McCabe's (2010) focus on four Maasai families, and Kawa's (2016) use of ethnography that examines "the stories behind the science" to uncover how people in the Brazilian Amazon both exert power and are powerless in their relationships with plants, soils, and forests. I especially align with Cepek (2012), who draws upon a particular charismatic leader to understand and communicate the broader political and practical landscape of environmentalism and Indigenous identity politics in Ecuador.

14. Neumann ([2005] 2014, 2). See also Paulson, Gezon, and Watts (2005); Peet, Robbins, and Watts (2010); Peet and Watts (2004); Perrault, Bridge, and McCarthy (2015).

15. Rocheleau (2008, 722). I take a poststructuralist emphasis on discourse, practice, and history in the production of natural and social systems, and also draw upon feminist political ecology's emphasis on how gender shapes the ecological landscape as well as economic and political relations, including how identity and multiplicities of meanings are exerted on specifically local contexts. See also Rocheleau, Thomas-Slayter, and Wangari ([1960] 2013).

16. Leff (2015). See also Bryant and Bailey (1997); Martinez-Alier (1991, 2014); Porto-Gonçalves and Leff (2015).

interconnections across difference can strategically reconfigure relations of power and perhaps produce promising new politics of people and place.

Pulling Back the Curtain on Practicing Anthropology

Anthropology is a nuanced combination of art and science, marked by a perpetual balance between intentional and rigorous data collection and measured analysis, on the one hand, and more ephemeral, organic, almost intuitive storytelling, on the other, that emerges from the dynamics of "being there" through long-term fieldwork. By coming to know people and their worlds, we produce knowledge that is deeply qualitative but also influenced both by anthropologists' theoretical perspectives as well as by how we choose to convey what we have come to know.[17]

I conducted fieldwork for this book steadily over the course of sixteen years. My most intensive field research spanned thirteen months between 2006 and 2009, and I have returned to the region nearly every year since, often with students and almost always with my family. This ongoing, longitudinal nature of my research has given me insight into the durability as well as the dynamism of leaders and social movements in this region. Most of my fieldwork has been (and continues to be) multisited, focusing on people and organizations in the cities of Ilhéus and Itabuna, the towns of Serra Grande and Itacaré, as well as smaller rural communities near these cities. I complement this deep ethnographic focus on activists in this region through additional research with leaders in Salvador, Brasília, and São Paulo to understand the perspectives of actors at state and national levels.

Doing anthropological research often involves opportunistically "following the leads," wherever they are, and I would set off for long days of research in my little white fuscinha, which came to mark my presence throughout the

---

17. See Carrithers et al. (1990) for an article on this duality within the field of anthropology, appropriately titled "On Ethnography without Tears." Though this text travels beyond anthropology, it is based on anthropological methods and invites other disciplines to discern and use aspects that are valuable in conducting deep qualitative research. McGranahan (2018) writes of ethnographic sensibility, something we *do* to understand people and their ways of living in the world and something we come to *know* through this process. See Bestemen (2013) for an interpretation of what "doing anthropology" involves, including concepts that mark the work of anthropologists from translation and critique to rupture and the role of imagination.

region. I took a mixed-methods approach to data collection. This approach drew upon semi-structured interviews, questionnaires, group interviews, and extensive participant observation at workshops, meetings, and events related to the social movements and the daily lives of social movement leaders.[18] Often "the meetings *are* the work," as an activist friend of mine observes.[19] Accordingly, I attended dozens of meetings, advisory councils, seminars, retreats, workshops, and training sessions that were organized and attended by environmentalists, farmers, quilombolas, and nativos. I also held interviews in people's homes, offices, and cafés, on beaches, in cars, traipsing through fields and forests, and waiting for appointments in public offices. All of my research and interactions were conducted exclusively in Portuguese, which I translated myself, and I relied on digital recordings and field notes to reconstruct these encounters.[20] I also collected organizational documents, publicity information, meeting minutes, and regional reports. Sometimes I brought the children along with me, hoping they might be entertained alongside my work. Other times I returned home late at night, missing dinner, showers, and important bedtime goodnights.

In the time since I first set foot in Southern Bahia nearly two decades ago, the boundaries between my life and "the field" have grown thin. Better cell phone coverage, instant messaging, and videoconferencing now enable me to communicate regularly with the people featured in this book. These new forms of communication have facilitated the participation of these leaders in their individual representation as well as in the collective aspects of knowledge production about their lives in Southern Bahia. Key characters in the book were invited to review a Portuguese draft of this manuscript and speak back to my interpretations, correcting me on everything from the smallest of details to broader themes and arguments.

----

18. My initial research included participant observation at fifty public meetings and events related to the different social movements and groups, eighty-five semi-structured interviews, twenty questionnaires, two group interviews, seven oral histories, and ongoing participant observation and informal interviews, which continued over the course of nearly two decades and still do to this day.

19. Moseley (2021).

20. I worked with two field assistants to help translate and administer questionnaires. An old-school foot pedal (which I wholeheartedly recommend!) greatly facilitated my transcription process by allowing me to listen to interviews in Portuguese and type them directly into English. I then used a coding program to draw out themes.

Over time, our forms of anthropological representation often change, evolve, and take on different forms of being depending upon the purpose they serve. While my doctoral dissertation provided the basis for this book, it has been liberatory to explore new forms of engaged and theoretically grounded work that allow me to foreground people's stories and lived experiences.[21] This style is intended to allow readers to immerse themselves, to "fully enter" Southern Bahia, while giving those who want more information the option to dig deeper through footnotes that complement the stories in the main text.

In my writing, I also wrestled with the issue of representation, namely whether to keep my characters' true identities. Adhering to the long-standing tradition in anthropology, I carefully chose pseudonyms for many people and had the primary characters choose alternate names that represented themselves. Something was unsettling about this, however. After consulting with other anthropologists and writers I trust, I placed this decision in the hands of the people themselves, giving them agency and power in deciding the format I would use. We discussed this together, and some of these activists even talked among themselves about it. All chose to use their original names and photos; they explained that they are public figures and have both personal as well as communal stories to tell. I won't forget the feeling, after painstakingly inserting the pseudonyms, of going back through the manuscript and diligently reinserting real names. I felt a profound sense of people "taking back their identities."

As all anthropologists do, I continually tacked between my role as a researcher and as a participant in life in this region. I was both apart from and a part of what I studied. While I attended meetings and conducted

---

21. While we write, we read. In the final stages of this manuscript, I was inspired by a brilliant new edited volume with a fitting title, *Writing Anthropology* (McGranahan 2020b). In chapters from this, Donna Goldstein cautions against "thin description" that flattens out the people we come to know, the characters and conduits for the messages we seek to convey. Carla Jones argues for using agitation as an impetus for engaged writing. Kim Fortun advocates for the value of storytelling about people and also about landscapes. And the ever-eloquent Kirin Narayan argues that ethnography matters "[f]or the discipline of paying attention, for learning from others, for becoming more responsibly aware of inequalities, for better understanding how the social forces causing suffering and how people might somehow find hope, and most generally for being perpetually pulled beyond the horizons of one's own taken-for-granted world" (quoted in McGranahan 2020a, 91). I am inspired, animated, and humbled by these insights, which have shaped my own writing.

interviews, I also walked in the forest, participated in community life, surfed the waves (or attempted to), ate the food that came from farmers' labor, enjoyed late-night live music and early-morning coffees. I also served, and continue to do so, on the board of a regional organization and published articles to raise awareness of the situations I was studying.[22]

Today, the people in this book continue as my partners in the dynamic and ongoing project of collective meaning-making and activism in Southern Bahia. When I get new ideas, I consult with them, as they do with me. When I bring students to the field, they are the experts I seek out to share their perspectives with newcomers to the region. These are my efforts to honor the myriad forms of knowledge production as well as to reflect how coming to know people in a place is an ongoing conversation and continual process of coproduction. As anthropologists conduct research and subsequently think and write about what we come to "know," we are challenged to demystify this process and bring into sharper relief what this entails on the part of the researcher. We can (and should) reflect more deeply upon the ways in which fieldwork and the knowledge that results from it is co-constructed by the researcher, in all of their positionality, as well as by the people being studied, who are both subjects of and agents in their stories. We should also recognize the power dynamics at play, the dualities, the "intersubjective creations" that constitute fieldwork and, in turn, are translated into final knowledge products that emerge out of these experiences.

Mothering in/of the Field

Anthropologists do not enter the field alone, with mechanical or dispassionate sensibilities. We carry with us our personal experiences, perceptions, biases, personalities, interests, theoretical training, and methodological strengths and weaknesses. We carry our positionality, our identities, our

---

22. I have long served on the Advisory Council of Instituto Nossa Ilhéus and engage in activism around Southern Bahia's development trajectory in my writing and teaching (see Scanlan Lyons 2009). I also work to coproduce collaborative and inclusive ethnographic practices, as well as products from these practices through co-authorship (Scanlan Lyons et al. 2018) and other forms of working with the people who serve as the experts in and subjects of anthropological research. See Hale (2006); Kirsch (2018); Lassiter (2005); Lassiter et al. (2005); and Rappaport (2008).

inhibitions and fears, as well as our hopes. We carry our past, present, and even future selves, which undoubtedly and indelibly shape how we see, hear, feel, understand, and, in turn, represent.[23]

I embarked on this project bearing particular life experiences that influenced, shaped, enhanced, and also sometimes inhibited my work. I began my career in anthropology with previous professional experiences in Brazil and also just as I was starting to have children.[24] Upon acceptance to graduate school in the spring of 1999, I walked into my new advisor's office and nervously announced I was pregnant with my first child. I feared this would indicate a lack of commitment, even though graduate school and my childbearing years happened to align. Her sage response has stuck with me since: "Go have that baby," she said without hesitation. "Being an anthropologist is a lifetime endeavor. Taking a semester off won't hurt you."[25]

I took her advice very much to heart and had three kids while doing my graduate coursework. Being in the field as a mother of three young children—Max, Ella, and Maya (ages five, three, and ten months, respectively, when I began my dissertation fieldwork)—shaped my daily practices and habits as anthropologist and mother and sometimes inverted my priorities. This

---

23. For decades anthropologists have been reflecting on power dynamics and the co-construction of knowledge through the fieldwork process (Clifford and Marcus 1986) and the power of feminist anthropology (Behar and Gordon 1995), a category in which I place myself. Nencel discusses how reflexivity and positionality are "second nature" to feminist anthropologists: "The text becomes a co-constructed space that reveals the interaction between the researchers' assumptions and the positionality and the voices, stories and experiences of the research subjects" (2014, 76).

24. In 1993, I lived in Manaus, Amazonas, for nearly a year and worked with the National Institute for Amazonian Research (INPA) and NGOs throughout the Amazon as part of my trajectory to earn a master's degree in international development. I also began my PhD already having professional work experience with U.S.-based NGOs, international consulting groups, a municipal government agency, and an international grant-making agency run by a global network of social-movement activists.

25. While I did fieldwork and completed my PhD, several of my mentors had children themselves. We became a small cohort of academic moms, sharing everything from childcare to fieldwork tips. Donna Goldstein, my friend and PhD advisor, who generously paved my academic path in Brazil—as well as one of the smartest women I know—once observed while embarking on fieldwork in Brazil with her daughter, "I've come up with a rough formula for fieldwork with kids: three weeks without a child in the field is about equal to six weeks *with* one." This gave us a good chuckle. Though our daughters are nearly adults now, we know all too well the complexity, fragility, and tenacity that accompany fieldwork with children in tow.

reality determined where I lived and for how long, as well as how I could conduct research. It also provided me with access and insights to the more personal lives of the people with whom I worked and, conversely, theirs into mine in ways that would have been unimaginable had I been alone. Mixing mothering with my research also allowed me to understand more deeply the lived experiences of social movement and activist leaders who similarly balance their professional calling with their personal lives.

I conducted just over a year of research with some combination of the children, usually with all three and, for a few months' time, with just my younger daughter, Maya. My husband, Jeff, who had less job flexibility, made the sixty-hour round-trip trek from Colorado to spend a week with the family every five weeks. This experience was exhausting and demanding, as well as exhilarating and fulfilling for us all.

Managing this chaos was not something I did naturally or deftly in the least. The vignettes and pictures of my family gazing at the camera represent how they observed me through this process. They were witnesses to my research, analysis, and writing about Bahia as well as contributors to their own versions of the stories of this place. To reflect aspects of this, I have drawn out certain elements to reflect how my scholarly research led me and my family on both individual and collective journeys that indelibly shaped our lives both in the past and also to this day.[26]

---

26. Analysis of fieldwork with families has grown markedly in the past few decades. For this reason, I titled this section "Mothering in/of the Field" to communicate our dual responsibility as parents and anthropologists to analyze and advocate for changing the field of anthropology in ways that allow people to bear both of these identities at the same time. One of my favorite articles that reflects on how gender, fieldwork, and theoretical insights converge is titled "'You Brought Your Baby to Base Camp?' Families and Field Sites" (Frohlick 2002). See also Brown and Casanova (2009); Castañeda (2019); Cupples and Kindon (2003); Drozdzewski and Robinson (2015); Flinn, Marshall, and Armstrong (1998); Jenkins (2020); and Korpela, Hirvi, and Tawah (2016).

And so, Dear Reader, we come to the end of "the story beneath the story," the footnotes that have hopefully served to elucidate and illuminate social movement activism in Southern Bahia. As Guiomar says, now that you have insights, go forth and tell others how people here are "running after," making their lives in this complex and dynamic and marvelous region of the world.

References

Adams, Cristina, Lucia Chamlian Munari, Nathalie Van Vliet, Rui Sergio Sereni Murrieta, Barbara Ann Piperata, Celia Futemma, Nelson Novaes Pedroso, Carolina Santos Taqueda, Mirella Abrahão Crevelaro, and Vânia Luísa Spressola-Prado. 2013. "Diversifying Incomes and Losing Landscape Complexity in Quilombola Shifting Cultivation Communities of the Atlantic Rainforest (Brazil)." *Human Ecology* 41 (1): 119–37.

Agrawal, Arun. 2005. *Environmentality: Technologies of Government and the Making of Subjects.* Durham, NC: Duke University Press.

Ahlborg, Helene, and Andrea Joslyn Nightingale. 2018. "Theorizing Power in Political Ecology: The Where of Power in Resource Governance Projects." *Journal of Political Ecology* 25 (1): 381–401.

Alger, Keith, and Marcellus Caldas. 1994. "The Declining Cocoa Economy and the Atlantic Forest of Southern Bahia, Brazil: Conservation Attitudes of Cocoa Planters." *Environmentalist* 14 (2): 107–19.

Almeida, Fernando Ozorio de, and Eduardo Góes Neves. 2015. "Evidências Arqueológicas para a Origem dos Tupi-Guarani no Leste da Amazônia." *Mana* 21 (3): 499–525.

Altieri, Miguel A., and Victor Manuel Toledo. 2011. "The Agroecological Revolution in Latin America: Rescuing Nature, Ensuring Food Sovereignty and Empowering Peasants." *The Journal of Peasant Studies* 38 (3): 587–612.

Alvarez, Sonia E., Evelnia Dagnino, and Arturo Escobar. [1998] 2018. "Introduction: The Cultural and the Political in Latin American Social Movements." In *Cultures of Politics, Politics of Cultures: Revisioning Latin American Social Movement,* edited by Sonia E. Alvarez, Evelina Dagnino, and Arturo Escobar, 1–30. London and New York: Routledge.

Alvim, Ronald, and P. K. R. Nair. 1986. "Combination of Cacao with Other Plantation Crops: An Agroforestry System in Southeast Bahia, Brazil." *Agroforestry Systems* 4 (1): 3–15.

Amado, Jorge. 1992. *The Golden Harvest.* New York: Avon Books.

Ames, Barry, and Margaret E. Keck. 1998. "The Politics of Sustainable Development Environmental Policy Making in Four Brazilian States." *Journal of Interamerican Studies and World Affairs* 39 (4): 1–40.

Andrade Souza, Jurema Machado de. 2017. "Remoções, Dispersões, e Reconfigurações Étnico-Territoriais entre os Pataxó Hãhãhãi." *Mediações-Revista de Ciências Sociais* 22 (2): 99–124.

Araujo, Quintino Reis de, Guilherme Amorim Homem de Abreu Loureiro, Virupax Chanabasappa Baligar, Dario Ahnert, José Claudio Faria, and Raul René Valle. 2019. "Cacao Quality Index for Cacao Agroecosystems in Bahia, Brazil." *International Journal of Food Properties* 22 (1): 1799–1814.

Azevedo, Andrea A., Raoni Rajão, Marcelo A. Costa, Marcelo CC Stabile, Marcia N. Macedo, Tiago NP Dos Reis, Ane Alencar, Britaldo S. Soares-Filho, and Rayane Pacheco. 2017. "Limits of Brazil's Forest Code as a Means to End Illegal Deforestation." *Proceedings of the National Academy of Sciences* 114 (29): 7653–58.

Banerjee, Onil, Alexander J. Macpherson, and Janaki Alavalapati. 2009. "Toward a Policy of Sustainable Forest Management in Brazil: A Historical Analysis." *The Journal of Environment and Development* 18 (2): 130–53.

Basso, Keith H. 2000. "Stalking with Stories." In *Schooling and the Symbolic Animal: Social and Cultural Dimensions of Education*, edited by Bradley A. U. Levinson, 41–52. Lanham, MD: Rowman & Littlefield.

Behar, Ruth, and Deborah A. Gordon, eds. 1995. *Women Writing Culture.* Berkeley: University of California Press.

Benjamin, Walter. 1969. *Illuminations.* New York: Schocken Books.

Besteman, Catherine. 2013. "Three Reflections on Public Anthropology." *Anthropology Today* 29 (6): 3–6.

Biehl, João, Byron Good, and Arthur Kleinman, eds. 2007. *Subjectivity: Ethnographic Investigations.* Berkeley: University of California Press.

Biersack, Aletta. 2006. "Reimagining Political Ecology: Culture/Power/History/Nature." In *Reimagining Political Ecology*, edited by Aletta Biersack and James B. Greenberg, 3–40. Durham, NC: Duke University Press.

Blaikie, Piers, and Harold Brookfield, eds. 2015. *Land Degradation and Society.* London and New York: Routledge.

Bogoni, Juliano André, José Salatiel Rodrigues Pires, Maurício Eduardo Graipel, Nivaldo Peroni, and Carlos A. Peres. 2018. "Wish You Were Here: How Defaunated Is the Atlantic Forest Biome of Its Medium- to Large-Bodied Mammal Fauna?" *PLOS One* 13 (9).

Brandon, Katrina, Gustavo A. B. Da Fonseca, Anthony B. Rylands, and José Maria Cardoso Da Silva. 2005. "Special Section: Brazilian Conservation: Challenges and Opportunities." *Conservation Biology* 19 (3): 595–600.

Branford, Sue, and Jan Rocha. 2002. *Cutting the Wire: The Story of the Landless Movement in Brazil*. London: Latin America Bureau.

Braun, Bruce. 2002. *The Intemperate Rainforest: Nature, Culture, and Power on Canada's West Coast*. Minneapolis: University of Minnesota Press.

Bright, Chris, and Radhika Sarin. 2003. *Venture Capitalism for a Tropical Forest: Cocoa in the Mata Atlântica*. Washington, DC: Worldwatch Institute.

Brockington, Dan. 2002. *Fortress Conservation: The Preservation of the Mkomazi Game Reserve, Tanzania*. Indianapolis: Indiana University Press.

Brockington, Daniel, and James Igoe. 2006. "Eviction for Conservation: A Global Overview." *Conservation and Society* 4 (3): 424–70.

Brosius, Peter J., Anna Lowenhaupt Tsing, and Charles Zerner, eds. 2005. *Communities and Conservation: Histories and Politics of Community-Based Natural Resource Management*. Lanham, MD: AltaMira Press.

Brown, Tamara Mose, and Erynn Masi de Casanova. 2009. "Mothers in the Field: How Motherhood Shapes Fieldwork and Researcher-Subject Relations." *Women's Studies Quarterly* 37 (3/4): 42–57.

Bryant, Raymond L., and Sinéad Bailey. 1997. *Third World Political Ecology*. London and New York: Routledge.

Burns, E. Bradford. 1993. *A History of Brazil*. New York: Columbia University Press.

Caldas, Marcellus M., and Stephen Perz. 2013. "Agro-Terrorism? The Causes and Consequences of the Appearance of Witch's Broom Disease in Cocoa Plantations of Southern Bahia, Brazil." *Geoforum* 47: 147–57.

Caldeira, Rute. 2009. "The Failed Marriage between Women and the Landless People's Movement (MST) in Brazil." *Journal of International Women's Studies* 10 (4): 237–58.

Caldeira, Teresa Pires do Rio. 1988. "The Art of Being Indirect: Talking about Politics in Brazil." *Cultural Anthropology* 3 (4): 444–54.

Campbell, Jeremy M. 2015. *Conjuring Property: Speculation and Environmental Futures in the Brazilian Amazon*. Seattle: University of Washington Press.

Carrithers, Michael, Chris Hahn, Thomas Hauschild, R. J. Thornton, and Paul A. Roth. 1990. "On Ethnography without Tears." *Current Anthropology* 31 (1): 53–58.

Carvalho, André Pereira de, and José Carlos Barbieri. 2013. "Inovações Socioambientais em Cadeias de Suprimento: Um Estudo de Caso sobre o Papel da Empresa Focal." *RAI Revista de Administração e Inovação* 10 (1): 232–56.

Cassano, Camila R., Götz Schroth, Deborah Faria, Jacques H. C. Delabie, and Lucio Bede. 2009. "Landscape and Farm Scale Management to Enhance Biodiversity Conservation in the Cocoa Producing Region of Southern Bahia, Brazil." *Biodiversity and Conservation* 18 (3): 577–603.

Castañeda, Angela. 2019. "Negotiating Fieldwork and Mothering" in *Travellin' Mama: Mothers, Mothering, and Travel*, edited by Charlotte Beyer, Janet MacLennan, Dorsía Smith Silva, and Marjoreie Tesser, 159–68. Bradford, Ontario: Demeter Press.

Castilho, Luciana C., Kristel M. De Vleeschouwer, E. J. Milner-Gulland, and Alexandre Schiavetti. 2019. "Hunting of Mammal Species in Protected Areas of the Southern Bahian Atlantic Forest, Brazil." *Oryx* 53 (4): 687–97.

Castree, Noel. 2007. "Neoliberal Ecologies." In *Neoliberal Environments: False Promises and Unnatural Consequences*, edited by Nik Heynen, James McCarthy, Scott Prudham, and Paul Robbins, 281–86. New York: Routledge.

Castree, Noel. 2008. "Neoliberalising Nature: The Logics of Deregulation and Reregulation." *Environment and Planning A* 40 (1): 131–52.

Castro, Hebe Maria Mattos de. 2006. "Terras de Quilombo: Campesinato, Memória do Cativeiro e Identidade Negra no Rio de Janeiro." In *Trabalho Livre, Trabalho Escravo: Brasil e Europa, Séculos XVIII e XIX*, edited by Douglas Cole Libby and Junia Ferreira Furtado, 415–36. São Paulo: Annablume.

Cehelsky, Marta. 2019. *Land Reform in Brazil: The Management of Social Change*. London and New York: Routledge.

Cepek, Michael. 2012. *A Future for Amazonia: Randy Borman and Cofán Environmental Politics*. Austin: University of Texas Press.

Chalhoub, Sidney. 2006. "The Politics of Silence: Race and Citizenship in Nineteenth-Century Brazil." *Slavery and Abolition* 27 (1): 73–87.

Chambers, Erve. 1997. "Introduction: Tourism's Mediators." *Tourism and Culture: An Applied Perspective*, edited by Erve Chambers, 1–12. New York: SUNY Press.

Chiapetti, Rita Jaqueline Nogueira. 2009. "Na Beleza do Lugar, o Rio das Contas indo . . . ao Mar." Doctoral thesis, Universidade Estadual Paulista, Instituto de Geociências e Ciências Exatas.

Chomitz, Kenneth M., Keith Alger, Timothy S. Thomas, Heloisa Orlando, and Paulo Vila Nova. 2005. "Opportunity Costs of Conservation in a Biodiversity Hotspot: The Case of Southern Bahia." *Environment and Development Economics* 10 (3): 293–312.

Clare, Nick, Victoria Habermehl, and Liz Mason-Deese. 2018. "Territories in Contestation: Relational Power in Latin America." *Territory, Politics, Governance* 6 (3): 302–21.

Clifford, James. 1997. *Routes: Travel and Translation in the Late Twentieth Century*. Cambridge, MA: Harvard University Press.

Clifford, James, and George E. Marcus, eds. 1986. *Writing Culture: The Poetics and Politics of Ethnography*. Berkeley: University of California Press.

Collins, John. 2008. "'But What If I Should Need to Defecate in Your Neighborhood, Madame?': Empire, Redemption, and the 'Tradition of the Oppressed' in a Brazilian World Heritage Site." *Cultural Anthropology* 23 (2): 279–328.

Collins, John F. 2015. *Revolt of the Saints: Memory and Redemption in the Twilight of Brazilian Racial Democracy*. Durham, NC: Duke University Press.

Comaroff, Jean, and John L. Comaroff. 2000. "Millennial Capitalism: First Thoughts on a Second Coming." *Public Culture* 12 (2): 291–343.

Conklin, Beth A., and Laura R. Graham. 1995. "The Shifting Middle Ground: Amazonian Indians and Eco-politics." *American Anthropologist* 97 (4): 695–710.

Connell, Raewyn, and Nour Dados. 2014. "Where in the World Does Neoliberalism Come From?" *Theory and Society* 43 (2): 117–38.

Connerton, Paul. 1989. *How Societies Remember*. Cambridge: Cambridge University Press.

Coronil, Fernando. 1996. "Beyond Occidentalism: Toward Nonimperial Geohistorical Categories." *Cultural Anthropology* 11 (1): 51–87.

Couto, Patrícia AB. 2007. "O Direito ao Lugar: Situacoes Processuals de Conflitos na Reconfiguracao Social e Territorial do Municipio de Itacaré, BA." Doctoral dissertation, Universidade Federal Fluminense. Niteroi, Rio de Janeiro.

Couto, Patrícia de Araújo Brandão. 2011. "Porto de Trás: Etnicidade, Turismo e Patrimonialização." *PASOS Revista de Turismo y Patrimonio Cultural* 9 (3): 19–30.

Crenshaw, Kimberlé. 1990. "Mapping the Margins: Intersectionality, Identity Politics, and Violence against Women of Color." *Stanford Law Review* 43 (1990): 1241–99.

Crespo, Samyra. 1993. *O Que o Brasileiro Pensa da Ecologia*. Brasilia: Museu de Astronomia e Ciências Afins/CNPq.

Crook, Larry. 2005. *Brazilian Music: Northeastern Traditions and the Heartbeat of a Modern Nation*. Santa Barbara and Denver: ABC-CLIO.

Cruikshank, Julie. 1998. *The Social Life of Stories: Narrative and Knowledge in the Yukon Territory*. Lincoln: University of Nebraska Press.

Cullen Jr., Laury, Keith Alger, and Denise M. Rambaldi. 2005. "Land Reform and Biodiversity Conservation in Brazil in the 1990s: Conflict and the Articulation of Mutual Interests." *Conservation Biology* 19 (3): 747–55.

Cummings, Jake. 2015. "Confronting Favela Chic: The Gentrification of Informal Settlements in Rio de Janeiro, Brazil." In *Global Gentrifications: Uneven Development and Displacement*, edited by Loretta Lees, Hyun Shin, and Ernesto López-Morales, 81–99. Bristol, UK: Policy Press.

Cupples, Julie, and Sara Kindon. 2003. "Far from Being 'Home Alone': The Dynamics of Accompanied Fieldwork." *Singapore Journal of Tropical Geography* 24 (2): 211–28.

Da Matta, Roberto. 1979. "Você Sabe Com Quem Está Falando." *Carnavais, Malandros e Heróis: Para Uma Sociologia do Dilema Brasileiro*. Rio de Janeiro: Rocco.

Das, Veena. 2008. "Violence, Gender, and Subjectivity." *Annual Review of Anthropology* 37: 283–99.

Davis, Joseph E., ed. 2012. *Stories of Change: Narrative and Social Movements*. Albany: SUNY Press.

Dean, Warren. 1997. *With Broadax and Firebrand: The Destruction of the Brazilian Atlantic Forest*. Berkeley: University of California Press.

Demeter, Paulo Roberto. 1996. *Combatendo o Desemprego na Regiao Cacaueira da Bahia: O Papel Dos Movimentos Sociais Populares*. Itabuna, Brasil: FASE (Federaçao de Órgaos Pra Assistência Social).

DeVore, Jonathan. 2015. "The Landless Invading the Landless: Participation, Coercion, and Agrarian Social Movements in the Cacao Lands of Southern Bahia, Brazil." *Journal of Peasant Studies* 42 (6): 1201–23.

DeVore, Jonathan. 2018. "Scattered Limbs: Capitalists, Kin, and Primitive Accumulation in Brazil's Cacao Lands, 1950s–1970s." *Journal of Latin American and Caribbean Anthropology* 23 (3): 496–520.

DeVore, Jonathan, Eric Hirsch, and Susan Paulson. 2019. "Conserver la Nature Humaine et Non Humaine un Curieux Cas de Conservation Conviviale au Brésil." *Anthropologie e Sociétés* 43 (3): 31–58.

Diegues, Antonio Carlos. 2014. "The Role of Ethnoscience in the Build-Up of Ethnoconservation as a New Approach to Nature Conservation in the Tropics: The Case of Brazil." *Revue d'ethnoécologie* 6: 1–16.

Doane, Molly. 2012. *Stealing Shining Rivers: Agrarian Conflict, Market Logic, and Conservation in a Mexican Forest.* Tucson: University of Arizona Press.

Dowie, Mark. 2011. *Conservation Refugees: The Hundred-Year Conflict between Global Conservation and Native Peoples.* Cambridge, MA: MIT Press.

Downey, Greg. 2005. *Learning Capoeira: Lessons in Cunning from an Afro-Brazilian Art.* Oxford: Oxford University Press.

Drozdzewski, Danielle, and Daniel F. Robinson. 2015. "Care-Work on Fieldwork: Taking Your Own Children into the Field." *Children's Geographies* 13 (3): 372–78.

Drummond, Jose Augusto, Jose Luiz de Andrade Franco, and Alessandra Bortoni Ninis. 2009. "Brazilian Federal Conservation Units: A Historical Overview of Their Creation and of Their Current Status." *Environment and History* 15 (4): 463–91.

Escallón, Maria Fernanda. 2019. "Rights, Inequality, and Afro-Descendant Heritage in Brazil." *Cultural Anthropology* 34 (3): 359–87.

Escobar, Arturo. 1992. "Culture, Practice and Politics: Anthropology and the Study of Social Movements." *Critique of Anthropology* 12 (40): 395–432.

Escobar, Arturo. 1998. "Whose Knowledge, Whose Nature? Biodiversity, Conservation, and the Political Ecology of Social Movements." *Journal of Political Ecology* 5 (1): 53–82.

Escobar, Arturo. 2001. "Culture Sits in Places: Reflections on Globalism and Subaltern Strategies of Localization." *Political Geography* 20 (2): 139–74.

Escobar, Arturo. 2005. "Imagining a Post-Development Era." In *Power of Development*, edited by Jonathan Crush, 211–27. London: Routledge.

Escobar, Arturo. 2008. *Territories of Difference: Place, Movements, Life, Redes.* Durham, NC: Duke University Press.

Escobar, Arturo. 2017. "Foreword." In *Beyond Civil Society: Activism, Participation, and Protest in Latin America*, edited by Sonia E. Alvarez, Jeffrey W. Rubin, Millie Thayer, Gianpaolo Balocchi, and Agustin Laó-Montes, ix–xii. Durham, NC: Duke University Press.

Fabian, Johannes. 1990. "Presence and Representation: The Other and Anthropological Writing." *Critical Inquiry* 16 (4): 753–72.

Farfán-Santos, Elizabeth. 2015. "'Fraudulent' Identities: The Politics of Defining Quilombo Descendants in Brazil." *The Journal of Latin American and Caribbean Anthropology* 20 (1): 110–32.

Farfán-Santos, Elizabeth. 2016. *Black Bodies, Black Rights: The Politics of Quilombolismo in Contemporary Brazil.* Austin: University of Texas Press.

Faria, Deborah, Jacques Hubert Charles Delabie, and Marcelo Henrique Dias. 2021. "The Hileia Baiana: An Assessment of Natural and Historical Aspects of the Land Use and Degradation of the Central Corridor of the Brazilian Atlantic Forest." In *The Atlantic Forest: History, Biodiversity, Threats and Opportunities of the Mega-Diverse Forest,* edited by Marcia C. M. Marques and Carlos E. V. Grelle, 63–90. Cham: Springer International Publishing.

Faria, Deborah, Rudi Ricardo Laps, Julio Baumgarten, and Maurício Cetra. 2006. "Bat and Bird Assemblages from Forests and Shade Cacao Plantations in Two Contrasting Landscapes in the Atlantic Forest of Southern Bahia, Brazil." *Biodiversity and Conservation* 15 (2): 587–612.

Faria, Deborah, Mateus Luís Barradas Paciencia, Marianna Dixo, Rudi Ricardo Laps, and Julio Baumgarten. 2007. "Ferns, Frogs, Lizards, Birds and Bats in Forest Fragments and Shade Cacao Plantations in Two Contrasting Landscapes in the Atlantic Forest, Brazil." *Biodiversity and Conservation* 16 (8): 2335–57.

Ferguson, James. 2010. "The Uses of Neoliberalism." *Antipode* 41: 166–84.

Ferreira, Simone Raquel Batista. 2006. "Quilombola 'Peasantiness' and Territoriality in North Espirito Santo." *GEOgraphia* 8 (16): 57–82.

Flachs, Andrew. 2019. *Cultivating Knowledge: Biotechnology, Sustainability, and the Human Cost of Cotton Capitalism in India.* Tucson: University of Arizona Press.

Flachs, Andrew, and Paul Richards. 2018. "Playing Development Roles: The Political Ecology of Performance in Agricultural Development." *Journal of Political Ecology* 25 (1): 638–46.

Fletcher, Robert. 2010. "Neoliberal Environmentality: Towards a Poststructuralist Political Ecology of the Conservation Debate." *Conservation and Society* 8 (3): 171–81.

Flinn, Juliana, Leslie Marshall, and Jocelyn Armstrong, eds. 1998. *Fieldwork and Families: Constructing New Models for Ethnographic Research.* Honolulu: University of Hawai'i Press.

Foley, Conor, ed. 2019. *In Spite of You: Bolsonaro and the New Brazilian Resistance.* New York: O/R Books.

Fonseca, Gustavo A. B. da. 2003. "Conservation Science and NGOs." *Conservation Biology* 17 (2): 345–47.

Fortun, Kim. 2020. "To Fieldwork, to Write." In *Writing Anthropology: Essays on Craft and Commitment,* edited by Carole McGranahan, 110–17. Durham, NC: Duke University Press.

Foucault, Michel. 1977. *Discipline and Punish: The Birth of the Prison.* Trans. Alan Sheridan. New York: Vintage Books.

French, Jan Hoffman. 2002. "Dancing for Land: Law-Making and Cultural Performance in Northeastern Brazil." *PoLAR* 25: 19–36.

French, Jan Hoffman. 2006. "Buried Alive: Imagining Africa in the Brazilian Northeast." *American Ethnologist* 33 (3): 340–60.

French, Jan Hoffman. 2009. *Legalizing Identities: Becoming Black or Indian in Brazil's Northeast*. Chapel Hill: University of North Carolina Press.

Freyre, Gilberto. (1946) 2013. *The Masters and the Slaves: A Study in the Development of Brazilian Civilization*. Trans. Samuel Putnam. New York: Knopf.

Frohlick, Susan E. 2002. "'You Brought Your Baby to Base Camp?' Families and Field Sites." *The Great Lakes Geographer* 9 (1): 49–58.

Galeano, Eduardo. (1973) 1997. *Open Veins of Latin America: Five Centuries of the Pillage of a Continent*. New York: NYU Press.

Garcia, Ana. 2019. "Brazil under Bolsonaro: Social Base, Agenda and Perspectives." *Journal of Global Faultlines* 6 (1): 62–69.

Garland, David. 2014. "What Is a 'History of the Present'? On Foucault's Genealogies and Their Critical Preconditions." *Punishment & Society* 16 (4): 365–84.

Geertz, Clifford. 1973. *The Interpretation of Cultures*. New York: Basic Books.

Giddens, Anthony. 1979. *Central Problems in Social Theory: Action, Structure, and Contradiction in Social Analysis*. Berkeley: University of California Press.

Goldman, Mara Jill. 2020. *Narrating Nature: Wildlife Conservation and Maasai Ways of Knowing*. Tucson: University of Arizona Press.

Goldman, Michael. 2005. *Imperial Nature: The World Bank and Struggles for Social Justice in the Age of Globalization*. New Haven, CT: Yale University Press.

Goldstein, Donna. 1999. "'Interracial' Sex and Racial Democracy in Brazil: Twin Concepts?" *American Anthropologist* 101 (3): 563–78.

Goldstein, Donna. 2003. *Laughter Out of Place: Race, Class, Violence, and Sexuality in a Rio Shantytown*. Berkeley: University of California Press.

Goldstein, Donna. 2020. "Beyond Thin Description: Biography, Theory, Ethnographic Writing." In *Writing Anthropology: Essays on Craft and Commitment*, edited by Carole McGranahan, 78–82. Durham, NC: Duke University Press.

Gramsci, Antonio. 2005. "The Intellectuals." *Contemporary Sociological Thought* 49: 60–69.

Guanziroli, Carlos, António Márcio Buainain, Gabriela Benatti, and Vahíd Shaikhzadeh Vahdat. 2019. "The Fate of Family Farming under the New Pattern of Agrarian Development in Brazil." In *Agricultural Development in Brazil: The Rise of a Global Agro-Food Power*, edited by Antonio M. Buainain, Rodrigo Lanna, and Zander Navarro, 174–88. London and New York: Routledge.

Gupta, Akhil, and James Ferguson. 1992. "Beyond 'Culture': Space, Identity, and the Politics of Difference." *Cultural Anthropology* 7 (1): 6–23.

Hale, Charles R. 1997. "Cultural Politics of Identity in Latin America." *Annual Review of Anthropology* 26 (1): 567–90.

Hale, Charles R. 2006. "Activist Research v. Cultural Critique: Indigenous Land Rights and the Contradictions of Politically Engaged Anthropology." *Cultural Anthropology* 21 (1): 96–120.

Hammond, John L. 2009. "Land Occupations, Violence, and the Politics of Agrarian Reform in Brazil." *Latin American Perspectives* 36 (4): 156–77.

Harper, Krista, and S. Ravi Rajan. 2007. "International Environmental Justice: Building the Natural Assets of the World's Poor." In *Reclaiming Nature: Environmental Justice and Ecological Restoration*, edited by James K. Boyce, Sunita Narain, and Elizabeth A. Stanton, 327–50. New York: Anthem Press.

Hartter, Joel, and Abraham Goldman. 2011. "Local Responses to a Forest Park in Western Uganda: Alternate Narratives on Fortress Conservation." *Oryx* 45 (1): 60–68.

Harvey, David. 2007. *A Brief History of Neoliberalism*. New York: Oxford University Press.

Hastings, Jesse G. 2011. "International Environmental NGOs and Conservation Science and Policy: A Case from Brazil." *Coastal Management* 39 (3): 317–35.

Herrmann, Julián Durazo. 2017. "Clientelism and State Violence in Subnational Democratic Consolidation in Bahia, Brazil." In *Violence in Latin America and the Caribbean: Subnational Structures, Institutions, and Clientelistic Networks*, edited by Tina Hilgers and Laura Macdonald, 211–29. Cambridge: Cambridge University Press.

Hess, David J., and Roberto Da Matta, eds. 1995. *The Brazilian Puzzle: Culture on the Borderlands of the Western World*. New York: Columbia University Press.

Hourneaux Jr., Flavio, Barbara Galleli, Dolores Gallardo-Vázquez, and M. Isabel Sánchez-Hernández. 2017. "Strategic Aspects in Sustainability Reporting in Oil & Gas Industry: The Comparative Case-Study of Brazilian Petrobras and Spanish Repsol." *Ecological Indicators* 72 (January): 203–14.

Htun, Mala. 2004. "From 'Racial Democracy' to Affirmative Action: Changing State Policy on Race in Brazil." *Latin American Research Review* 39 (1): 60–89.

Igoe, Jim. 2017. *The Nature of Spectacle: On Images, Money, and Conserving Capitalism*. Tucson: University of Arizona Press.

Jenkins, Katy. 2020. "Academic Motherhood and Fieldwork: Juggling Time, Emotions, and Competing Demands." *Transactions of the Institute of British Geographers* 45 (3): 693–704.

Johns, Norman D. 1999. "Conservation in Brazil's Chocolate Forest: The Unlikely Persistence of the Traditional Cocoa Agroecosystem." *Environmental Management* 23 (1): 31–47.

Johnston, Hank, and Bert Klandermans, eds. 2013. *Social Movements and Culture*. London and New York: Routledge.

Jones, Carla. 2020. "A Case for Agitation: On Affect and Writing." In *Writing Anthropology: Essays on Craft and Commitment*, edited by Carole McGranahan, 145–48. Durham, NC: Duke University Press.

Kamino, Luciana Hiromi Yoshino, Eric Oliveira Pereira, and Flávio Fonseca do Carmo. 2020. "Conservation Paradox: Large-Scale Mining Waste in Protected Areas in Two Global Hotspots, Southeastern Brazil." *Ambio* 49 (10): 1629–38.

Kawa, Nicholas C. 2016. *Amazonia in the Anthropocene: People, Soils, Plants, Forests*. Austin: University of Texas Press.

Keck, Margaret E. 1995. "Social Equity and Environmental Politics in Brazil: Lessons from the Rubber Tappers of Acre." *Comparative Politics* 27 (4): 409–24.

Kempf, Victor. 2019. "Exodus from the Political: Workerist Conceptions of Radical Resistance." In *Rule and Resistance beyond the Nation-State: Contestation, Escalation, Exit,* edited by Felix Anderl, Christopher Daase, Nicole Deitelhoff, Victor Kempf, Jannik Pfister, and Philip Wallmeier, 257–78. London and New York: Rowman & Littlefield.

Kirsch, Stuart. 2018. *Engaged Anthropology: Politics beyond the Text.* Berkeley: University of California Press.

Korpela, Mari, Laura Hirvi, and Sanna Tawah. 2016. "Not Alone: Doing Fieldwork in the Company of Family Members." *Suomen Antropologi: Journal of the Finnish Anthropological Society* 41 (3): 3–20.

Koslinski, Mariane Campelo, and Elisa P. Reis. 2009. "Transnational and Domestic Relations of NGOs in Brazil." *World Development* 37 (3): 714–25.

Kraay, Hendrik. 1998. "Reconsidering Recruitment in Imperial Brazil." *The Americas* 55 (1): 1–33.

Kraay, Hendrik. 2016. *Afro-Brazilian Culture and Politics: Bahia, 1790s to 1990s.* London and New York: Routledge.

Kurtz, Esther Viola. 2020. "Guerreira Tactics: Women Warriors' Sonic Practices of Refusal in Capoeira Angola." *Women and Music: A Journal of Gender and Culture* 24 (1): 71–95.

Laffitte, Stefania Capone. 2010. *Searching for Africa in Brazil: Power and Tradition in Candomblé.* Durham, NC: Duke University Press.

Lamberti, Filippo. 2017. "Economic History of Cocoa in Southern Bahia: Its Role on Economy, Society and Culture." Doctoral dissertation, Fundação Getulio Vargas.

Landau, Elena Charlotte, Roberta Kelly da Cruz, André Hirsch, Fernando Martins Pimenta, and Daniel Pereira Guimarães. 2012. *Variação Geográfica do Tamanho dos Módulos Fiscais no Brasil.* Brasilia: Embrapa Milho e Sorgo-Documentos (INFOTECA-E).

Lassiter, Luke Eric. 2005. *The Chicago Guide to Collaborative Ethnography.* Chicago: University of Chicago Press.

Lassiter, Luke Eric, Samuel R. Cook, Les Field, Sjoerd R. Jaarsma, James L. Peacock, Deborah Rose, and Brian Street. 2005. "Collaborative Ethnography and Public Anthropology." *Current Anthropology* 46 (1): 83–106.

Lavoie, Anna, and Christian Brannstrom. 2019. "Assembling a Marine Extractive Reserve: The Case of the Cassurubá RESEX in Brazil." *Journal of Latin American Geography* 18 (2): 120–51.

Leal, Carlos Galindo, and Ibsen de Gusmão Câmara, eds. 2003. *The Atlantic Forest of South America: Biodiversity Status, Threats, and Outlook.* Washington, DC: Island Press.

Leff, Enrique. 2015. "Political Ecology: A Latin American Perspective." *Desenvolvimento e Meio Ambiente* 35: 29–64.

Leite, Ilka Boaventura. 2000. "Os Quilombos no Brasil: Questões Conceituais e Normativas." *Etnográfica* 4 (2): 333–54.

Leite, Ilka Boaventura. 2012. "The Transhistorical, Juridical-Formal, and Post-Utopian Quilombo." In *New Approaches to Resistance in Brazil and Mexico*, edited by John Gledhill and Patience A. Schell, 250–68. Durham, NC: Duke University Press.

Leite, Ilka Boaventura. 2015. "The Brazilian Quilombo: 'Race,' Community and Land in Space and Time." *Journal of Peasant Studies* 42 (6): 1225–40.

Li, Tanya Murray. 2000. "Articulating Indigenous Identity in Indonesia: Resource Politics and the Tribal Slot." *Comparative Studies in Society and History* 42 (1): 149–79.

Linhares, Luiz Fernando do Rosário. 2004. "Kilombos of Brazil: Identity and Land Entitlement." *Journal of Black Studies* 34 (6): 817–37.

Logan, Shannon, and Gerda R. Wekerle. 2008. "Neoliberalizing Environmental Governance? Land Trusts, Private Conservation and Nature on the Oak Ridges Moraine." *Geoforum* 39 (6): 2097–2108.

Loureiro, Pedro Mendes, and Alfredo Saad-Filho. 2019. "The Limits of Pragmatism: The Rise and Fall of the Brazilian Workers' Party (2002–2016)." *Latin American Perspectives* 46 (1): 66–84.

Lovell, Peggy A. 1994. "Race, Gender, and Development in Brazil." *Latin American Research Review* 29 (3): 7–35.

Luna, Francisco Vidal, and Herbert S. Klein. 2004. "Slave Economy and Society in Minas Gerais and São Paulo, Brazil in 1830." *Journal of Latin American Studies* 36 (1): 1–28.

Lyons, Kristina M. 2020. *Vital Decomposition: Soil Practitioners and Life Politics*. Durham, NC: Duke University Press.

Mahony, Mary Ann. 2001. "A Past to Do Justice to the Present: Collective Memory, Historical Representation, and Rule in Bahia's Cacao Area." In *Reclaiming the Political in Latin American History*, edited by Gilbert M. Joseph and Emily S. Rosenberg, 485–525. Durham, NC: Duke University Press.

Mahony, Mary Ann. 2016. "Afro-Brazilians, Land Reform, and the Question of Social Mobility in Southern Bahia, 1880–1920." In *Afro-Brazilian Culture and Politics: Bahia, 1790s to 1990s*, edited by Henrik Kraay, 90–91. New York: Routledge.

Mainwaring, Scott. 1999. *Rethinking Party Systems in the Third Wave of Democratization: The Case of Brazil*. Redwood City, CA: Stanford University Press.

Martinez-Alier, Joan. 1991. "Ecology and the Poor: A Neglected Dimension of Latin American History." *Journal of Latin American Studies* 23 (3): 621–39.

Martinez-Alier, Joan. 2014. "The Environmentalism of the Poor." *Geoforum* 54: 239–41.

Martínez-Reyes, José E. 2016. *Moral Ecology of a Forest: The Nature Industry and Maya Post-Conservation*. Tucson: University of Arizona Press.

Martini, Adriana Maria Zanforlin, Pedro Fiaschi, André M. Amorim, and José Lima Da Paixão. 2007. "A Hot-Point within a Hot-Spot: A High Diversity Site in Brazil's Atlantic Forest." *Biodiversity and Conservation* 16 (11): 3111–28.

Mattos, Hebe. 2005. "'Remanescentes de Quilombos': Memory of Slavery, Historical Justice, and Citizenship in Contemporary Brazil." In *Proceedings of the 7th Annual Gilder Lehrmann Center International Conference, Yale University*, October 27, 2005, 27–29.

McAdam, Doug. 1994. "Culture and Social Movements." In *New Social Movements: From Ideology to Identity*, edited by Enrique Laraña, Hank Johnston, and Joseph R. Gusfield, 36–57. Philadelphia: Temple University Press.

McCabe, J. Terrence. 2010. *Cattle Bring Us to Our Enemies: Turkana Ecology, Politics, and Raiding in a Disequilibrium System*. Ann Arbor: University of Michigan Press.

McCarthy, James. 2002. "First World Political Ecology: Lessons from the Wise Use Movement." *Environment and Planning A* 34 (7): 1281–1302.

McGranahan, Carole. 2005. "Truth, Fear, and Lies: Exile Politics and Arrested Histories of the Tibetan Resistance." *Cultural Anthropology* 20 (4): 570–600.

McGranahan, Carole. 2018. "Ethnography beyond Method: The Importance of an Ethnographic Sensibility." *Sites: A Journal of Social Anthropology and Cultural Studies* 15 (1).

McGranahan, Carole. 2020a. "Ethnographic Writing with Kirin Narayan." In *Writing Anthropology: Essays on Craft and Commitment*, edited by Carole McGranahan, 87–92. Durham, NC: Duke University Press.

McGranahan, Carole. 2020b. *Writing Anthropology: Essays on Craft and Commitment*. Durham, NC: Duke University Press.

Mitchell, Sean. 2017. *Constellations of Inequality*. Chicago: University of Chicago Press.

Mittermeier, Russell A., Gustavo a. B. Da Fonseca, Anthony B. Rylands, and Katrina Brandon. 2005. "A Brief History of Biodiversity Conservation in Brazil." *Conservation Biology* 19 (3): 601–7.

Mohanty, Chandra Talpade. 2003. *Feminism without Borders: Decolonizing Theory, Practicing Solidarity*. Durham, NC: Duke University Press.

Moore, Donald S., Anand Pandian, and Jake Kosek. 2003. "Introduction: The Cultural Politics of Race and Nature: Terrains of Power and Pracitce." In *Race, Nature, and the Politics of Difference*, edited by Donald S. Moore, Jake Kosek, and Anand Pandian, 1–70. Durham, NC: Duke University Press.

Morellato, L. Patrícia C., and Célio FB Haddad. 2000. "Introduction: The Brazilian Atlantic Forest." *Biotropica* 32 (4b): 786–92.

Mori, Scott A., and Luiz Alberto Mattos Silva. 1979. "The Herbarium of the 'Centro de Pesquisas do Cacau' at Itabuna, Brazil." *Brittonia* 31 (2): 177–96.

Moseley, Matthew L. 2021. *Ignition: Superior Communication Strategies for Creating Stronger Connections*. CRC Press.

Motta, Renata. 2016. "Global Capitalism and the Nation State in the Struggles over GM Crops in Brazil." *Journal of Agrarian Change* 16 (4): 720–27.

Moura, Rodrigo Leão De, Carolina Viviana Minte-Vera, Isabela Baleeiro Curado, Ronaldo Bastos Francini-Filho, Hélio De Castro Lima Rodrigues, Guilherme Fraga Dutra, Diego Corrêa Alves, and Francisco José Bezerra Souto. 2009. "Challenges and Prospects of Fisheries Co-Management under a Marine Extractive Reserve Framework in Northeastern Brazil." *Coastal Management* 37 (6): 617–32.

Myers, Norman. 1988. "Threatened Biotas: 'Hot Spots' in Tropical Forests." *Environmentalist* 8 (30): 187–208.

Myers, Norman, Russell A. Mittermeier, Cristina G. Mittermeier, Gustavo A B da Fonseca, and Jennifer Kent. 2000. "Biodiversity Hotspots for Conservation Priorities." *Nature* 403 (6772): 853–58.

Nencel, Lorraine. 2014. "Situating Reflexivity: Voices, Positionalities and Representations in Feminist Ethnographic Texts." *Women's Studies International Forum* 43: 75–83.

Neumann, Roderick P. (2005) 2014. *Making Political Ecology*. London and New York: Routledge.

Nogueira, Renata Fernandes, Iris Roitman, Fabrício Alvim Carvalho, Gustavo Taboada Soldati, and Tamiel Khan Baiocchi Jacobson. 2019. "Challenges for Agroecological and Organic Management of Cabruca Cocoa Agroecosystems in Three Rural Settlements in South Bahia, Brazil: Perceptions from Local Actors." *Agroforestry Systems* 93 (5): 1961–72.

Nugent, Stephen. 1997. "The Coordinates of Identity in Amazonia: At Play in the Fields of Culture." *Critique of Anthropology* 17 (1): 33–51.

Offen, Karl H. 2003. "The Territorial Turn: Making Black Territories in Pacific Colombia." *Journal of Latin American Geography* 2 (1): 43–73.

Oliveira, Athila Leandro de, Marcondes Geraldo Coelho Junior, Dalmo Arantes Barros, Alexander Silva de Resende, Jerônimo Boelsums Barreto Sansevero, Luis Antônio Coimbra Borges, Vanessa Maria Basso, and Sergio Miana de Faria. 2020. "Revisiting the Concept of 'Fiscal Modules': Implications for Restoration and Conservation Programs in Brazil." *Land Use Policy* 99: 104978.

Oliveira, Elton Silva. 2007. "Impactos Socioambiantales y Económicos del Turismo y sus Repercusiones en el Desarrollo Local: El Caso del Municipio de Itacaré Bahia." *Interações (Campo Grande)* 8 (2): 193–202.

Oliveira, José Antonio Puppim de. 2005. "Enforcing Protected Area Guidelines in Brazil: What Explains Participation in the Implementation Process?" *Journal of Planning Education and Research* 24 (4): 420–36.

Ondetti, Gabriel. 2008. "Up and Down with the Agrarian Question: Issue Attention and Land Reform in Contemporary Brazil." *Politics & Policy* 36 (4): 510–41.

Ondetti, Gabriel. 2010. *Land, Protest, and Politics: The Landless Movement and the Struggle for Agrarian Reform in Brazil*. University Park: Penn State University Press.

Ondetti, Gabriel. 2016. "The Social Function of Property, Land Rights and Social Welfare in Brazil." *Land Use Policy* 50 (January): 29–37.

Ong, Aihwa. 2006. *Neoliberalism as Exception: Mutations in Citizenship and Sovereignty*. Durham, NC: Duke University Press.

Onsrud, Hazel, Susan Elizabeth Nichols, and Silvane Paixao. 2006. *Women and Land Reform in Brazil*. Fredericton: Department of Geodesy and Geomatics Engineering, University of New Brunswick. http://www2.unb.ca/gga/Pubs/TR239.pdf.

Ortner, Sherry B. 2005. "Subjectivity and Cultural Critique." *Anthropological Theory* 5 (1): 31–52.

Page, Joseph A. 1996. *The Brazilians*. Lebanon, IN: Da Capo Press.

Palmer, Christian T. 2014. "Tourism, Changing Architectural Styles, and the Production of Place in Itacaré, Bahia, Brazil." *Journal of Tourism and Cultural Change* 12 (4): 349–63.

Palmer, Christian. 2017. "The Brazilian Hawaii: Surf Culture, Tourism, and the Construction of Place." *Global Ethnographic* 4: 1–9.

Partelow, Stefan, Marion Glaser, Sofía Solano Arce, Roberta Sá Leitão Barboza, and Achim Schlüter. 2018. "Mangroves, Fishers, and the Struggle for Adaptive Comanagement: Applying the Social-Ecological Systems Framework to a Marine Extractive Reserve (RESEX) in Brazil." *Ecology and Society* 23 (3): Art. 19.

Paulson, Susan, Lisa L. Gezon, and Michael Watts. 2005. "Politics, Ecologies, Genealogies." In *Political Ecology across Spaces, Scales, and Social Groups*, edited by Susan Paulson and Lisa L. Gezon, 17–37. New Brunswick, NJ: Rutgers University Press.

Pearce, Jenny. 2010. "Is Social Change Fundable? NGOs and Theories and Practices of Social Change." *Development in Practice* 20 (6): 621–35.

Peet, Richard, Paul Robbins, and Michael Watts. 2010. *Global Political Ecology*. London and New York: Routledge.

Peet, Richard, and Michael Watts, eds. 2004. *Liberation Ecologies: Environment, Development, Social Movements*, 2nd ed. London and New York: Routledge.

Peluso, Nancy Lee, and Michael R. Watts. 2001. *Violent Environments*. Ithaca, NY: Cornell University Press.

Perlman, Janice. 2010. *Favela: Four Decades of Living on the Edge in Rio de Janeiro*. Oxford: Oxford University Press.

Perreault, Tom, Gavin Bridge, and James McCarthy, eds. 2015. *The Routledge Handbook of Political Ecology*. London and New York: Routledge.

Perry, Keisha-Khan Y. 2016. "Geographies of Power: Black Women Mobilizing Intersectionality in Brazil." *Meridians* 14 (1): 94–120.

Peterson, Richard B., Diane Russell, Paige West, and J. Peter Brosius. 2010. "Seeing (and Doing) Conservation through Cultural Lenses." *Environmental Management* 45 (1): 5–18.

Petras, James. 1997. "Imperialism and NGOs in Latin America." *Monthly Review* 49 (7): 10.

Piasentin, Flora Bonazzi, and Carlos Hiroo Saito. 2012. "Caracterização do Cultivo de Cacau na Região Econômica Litoral Sul, Sudeste Da Bahia." *Estudo & Debate*, Lajeado 19 (2): 63–80.

Piotto, Daniel, Florencia Montagnini, Wayt Thomas, Mark Ashton, and Chadwick Oliver. 2009. "Forest Recovery after Swidden Cultivation across a 40-Year Chronosequence in the Atlantic Forest of Southern Bahia, Brazil." *Plant Ecology* 205 (2): 261–72.

Plummer, Dawn, and Betsy Ranum. 2002. "Brazil's Landless Workers Movement—Movimiento de Trabajadores Rurales Sem Terra—MST." *Social Policy* 33 (1): 18–23.

Porto-Gonçalves, Carlos Walter, and Enrique Leff. 2015. "Political Ecology in Latin America: The Social Re-Appropriation of Nature, the Reinvention of Territories and the Construction of an Environmental Rationality." *Desenvolvimento e Meio Ambiente* 35 (1): 65–88.

Price, Richard, ed. 1996. *Maroon Societies: Rebel Slave Communities in the Americas*. Baltimore: Johns Hopkins University Press.

Price, Richard. 1998. "Scrapping Maroon History: Brazil's Promise, Suriname's Shame." *New West Indian Guide/Nieuwe West-Indische Gids* 72 (3–4): 233–55.

Quadros Jr., Antonio Carlos de, and Catia Mary Volp. 2005. "Forró Universitário: A Tradução do Forró Nordestino no Sudeste Brasileiro." *Motriz: Journal of Physical Education UNESP* 11 (2):117–20.

Rabinow, Paul, George E. Marcus, James D. Faubion, and Tobias Rees. 2008. *Designs for an Anthropology of the Contemporary*. Durham, NC: Duke University Press.

Ramos, Alcida Rita. 1994. "The Hyperreal Indian." *Critique of Anthropology* 14 (2): 153–71.

Rappaport, Joanne. 2008. "Beyond Participant Observation: Collaborative Ethnography as Theoretical Innovation." *Collaborative Anthropologies* 1 (1): 1–31.

Reid, John, and Wilson Cabral de Sousa Jr. 2005. "Infrastructure and Conservation Policy in Brazil." *Conservation Biology* 19 (3): 740–46.

Reis, João José. 1996. "Escravos e Coiteiros no Quilombo do Oitizeiro—Bahia, 1806." In *Liberdade por um Fio: A História dos Quilombos no Brasil*, edited by João José Reis and Flávio dos Santos Gomez, 332–72. São Paulo: Companhía das Letras.

Rezende, Camila Linhares, Fabio Rubio Scarano, Eduardo Delgado Assad, Carlos Alfredo Joly, Jean Metzger, Bernardo Baeta Neves Strassburg, Marcelo Tabarelli, Gustavo Alberto Fonseca, and Russell A. Mittermeier. 2018. "From Hotspot to Hopespot: An Opportunity for the Brazilian Atlantic Forest." *Perspectives in Ecology and Conservation* 16 (4): 208–14.

Ribeiro, Milton Cezar, Alexandre Camargo Martensen, Jean Paul Metzger, Marcelo Tabarelli, Fábio Scarano, and Marie-Josee Fortin. 2011. "The Brazilian Atlantic Forest: A Shrinking Biodiversity Hotspot." In *Biodiversity Hotspots*, edited by Frank E. Zachos and Jan Christian Habel, 405–34. Berlin: Springer.

Robles, Wilder. 2018. "Revisiting Agrarian Reform in Brazil, 1985–2016." *Journal of Developing Societies* 34 (1): 1–34.

Rocheleau, Dianne E. 2008. "Political Ecology in the Key of Policy: From Chains of Explanation to Webs of Relation." *Geoforum* 39 (2): 716–27.

Rocheleau, Dianne, Barbara Thomas-Slayter, and Esther Wangari, eds. (1960) 2013. *Feminist Political Ecology: Global Issues and Local Experience*. London and New York: Routledge.

Rogers, Thomas D. 2010. *The Deepest Wounds: A Labor and Environmental History of Sugar in Northeast Brazil*. Chapel Hill: University of North Carolina Press.

Roland, Lorecia Kaifa. 2011. *Cuban Color in Tourism and La Lucha: An Ethnography of Racial Meanings*. Oxford: Oxford University Press.

Rolim, Samir G., and Adriano G. Chiarello. 2004. "Slow Death of Atlantic Forest Trees in Cocoa Agroforestry in Southeastern Brazil." *Biodiversity and Conservation* 13 (14): 2679–94.

Roseberry, William. 1997. "Marx and Anthropology." *Annual Review of Anthropology* 26 (1): 25–46.

Rosset, Peter, Raj Patel, and Michael Courville, eds. 2006. *Promised Land: Competing Visions of Agrarian Reform.* Oakland, CA: Food First Books.

Roth-Gordon, Jennifer. 2007. "Youth, Slang, and Pragmatic Expressions: Examples from Brazilian Portuguese1." *Journal of Sociolinguistics* 11 (3): 322–45.

Roth-Gordon, Jennifer. 2016. *Race and the Brazilian Body: Blackness, Whiteness, and Everyday Language in Rio de Janeiro.* Oakland: University of California Press.

Saatchi, Sassan, Donat Agosti, Keith Alger, Jacques Delabie, and John Musinsky. 2001. "Examining Fragmentation and Loss of Primary Forest in the Southern Bahian Atlantic Forest of Brazil with Radar Imagery." *Conservation Biology* 15 (4): 867–75.

Said, Edward. 1978. *Orientalism.* New York: Vintage.

Salamon, Lester M. 1994. "The Rise of the Nonprofit Sector." *Foreign Affairs* (July/August): 109–22.

Salman, Ton, and Willem Assies. 2017. "Anthropology and the Study of Social Movements." In *Handbook of Social Movements across Disciplines,* edited by Conny Roggeband and Bert Klandermans, 57–101. Cham: Springer.

Sambuichi, Regina Helena Rosa, and Mundayatan Haridasan. 2007. "Recovery of Species Richness and Conservation of Native Atlantic Forest Trees in the Cacao Plantations of Southern Bahia in Brazil." *Biodiversity and Conservation* 16 (13): 3681–3701.

Santilli, Juliana. 2005. *Socioambientalismo e Novos Direitos-Proteção Jurídica à Diversidade Biológica e Cultural.* São Paulo: Editora Peirópolis Ltda.

Santos, Flávio Gonçalves dos. 2019. "Dois Caminhos: Porto e as Opções de Trabalho no Sul da Bahia entre 1872 e 1940." *Almanack* 21 (April): 205–38.

Santos, Thiago Coelho, and Leila Oliveira Santos. 2021. "Diagnóstico Ambiental e os Conflitos Socioambientais da Zona Costeira do Município de Ilhéus—Bahia." *Meio Ambiente (Brasil)* 3 (1): 80–88.

Sauer, Sérgio. 2006. "The World Bank's Market-Based Land Reform in Brazil." In *Promised Land: Competing Visions of Agrarian Reform,* edited by Michael Courville, Peter Rosset, and Raj Patel, 177–91. Oakland, CA: Food First Books.

Scanlan Lyons, Colleen M. 2009. "Battle in Bahia: A New Port Faces Growing Resistance." *NACLA Report on the Americas* 42 (5): 30–34.

Scanlan Lyons, Colleen M. 2011. "Spaces of Silence and Efforts toward Voice: Negotiation and Power among 'Quilombo' Communities in Southern Bahia Brazil." *Afro-Hispanic Review* 30 (2): 115–32.

Scanlan Lyons, Colleen M., Maria DiGiano, Jason Gray, Javier Kinney, Magaly Medeiros, and Francisca Oliveira de Lima Costa. 2018. "Negotiating Climate Justice at the Subnational Scale: Challenges and Collaborations between Indigenous Peoples and Subnational Governments." In *Routledge Handbook of Climate Justice,* edited by Tanseen Jafry, 431–48. London and New York: Routledge.

Scheper-Hughes, Nancy. 1992. *Death without Weeping: The Violence of Everyday Life in Brazil.* Berkeley: University of California Press.

Scherer-Warren, Ilse. 2006. "Das Mobilizações as Redes de Movimentos Sociais." *Sociedade e Estado* 21 (1): 109–30.

Schmink, Marianne. 1982. "Land Conflicts in Amazonia." *American Ethnologist* 9 (2): 341–57.

Schroth, Götz, Deborah Faria, Marcelo Araujo, Lucio Bede, Sunshine A. Van Bael, Camila R. Cassano, Leonardo C. Oliveira, and Jacques H. C. Delabie. 2011. "Conservation in Tropical Landscape Mosaics: The Case of the Cacao Landscape of Southern Bahia, Brazil." *Biodiversity and Conservation* 20 (8): 1635–54.

Schwartz, Stuart B. 1985. *Sugar Plantations in the Formation of Brazilian Society: Bahia, 1550–1835.* Cambridge: Cambridge University Press.

Schwartz, Stuart B. 1996. *Slaves, Peasants, and Rebels: Reconsidering Brazilian Slavery.* Champaign: University of Illinois Press.

Scott, James C. 1998. *Seeing like a State: How Certain Schemes to Improve the Human Condition Have Failed.* New Haven, CT: Yale University Press.

Shore, Edward. 2017. "Geographies of Resistance: *Quilombos*, Afro-Descendants, and the Struggle for Land and Environmental Justice in Brazil's Atlantic Forest." *Afro-Hispanic Review* 36 (1): 58–78.

Silva dos Santos, Jade, Joanison Vicente dos Santos Teixeira, Deyna Hulda Arêas Guanaes, Wesley Duarte da Rocha, and Alexandre Schiavetti. 2020. "Conflicts between Humans and Wild Animals in and Surrounding Protected Area (Bahia, Brazil): An Ethnozoological Approach." *Ethnobiology and Conservation* 9: 1–23.

Skidmore, Thomas E. 2009. *Brazil: Five Centuries of Change.* 2nd ed. Oxford: Oxford University Press.

Soares-Filho, Britaldo, Raoni Rajão, Marcia Macedo, Arnaldo Carneiro, William Costa, Michael Coe, Hermann Rodrigues, and Ane Alencar. 2014. "Cracking Brazil's Forest Code." *Science* 344 (6182): 363–64.

Sodikoff, Genese. 2008. "Forest Conservation and Low Wage Labour." In *Greening the Great Red Island: Madagascar in Nature and Culture*, edited by Jeffrey Charles Kaufmann, 69–89. Pretoria: Africa Institute of South Africa.

Sodikoff, Genese. 2009. "The Low-wage Conservationist: Biodiversity and Perversities of Value in Madagascar." *American Anthropologist* 111 (4): 443–55.

SOS Mata Atlântica. 2019. "Atlas dos Remanescentes Florestais da Mata Atlântica." *Relatório Técnico 2019.* https://cms.sosma.org.br/wp-content/uploads/2021/05/ SOSMA_Atlas-da-Mata-Atlantica_2019-2020.pdf (accessed November 14, 2021).

Soterroni, Aline C., Aline Mosnier, Alexandre X. Y. Carvalho, Gilberto Câmara, Michael Obersteiner, Pedro R. Andrade, Ricardo C. Souza, et al. 2018. "Future Environmental and Agricultural Impacts of Brazil's Forest Code." *Environmental Research Letters* 13 (7): 074021.

Souza Martins, José de. 2002. "Representing the Peasantry? Struggles for/about Land in Brazil." *Journal of Peasant Studies* 29 (3–4): 300–35.

Stronzake, Judite, and Wendy Wolford. 2016. "Brazil's Landless Workers Rise Up." *Dissent* 63 (2): 48–55.

Sundberg, Juanita. 2006. "Conservation Encounters: Transculturation in the 'Contact Zones' of Empire." *Cultural Geographies* 13 (2): 239–65.

Swidler, Ann. 2000. "Cultural Power and Social Movements." In *Culture and Politics*, edited by Lane Crothers and Charles Lockhart, 269–83. Berlin: Springer.

Tabarelli, Marcelo, Luiz Paulo Pinto, Jose MC Silva, Marcia Hirota, and Lucio Bede. 2005. "Challenges and Opportunities for Biodiversity Conservation in the Brazilian Atlantic Forest." *Conservation Biology* 19 (3): 695–700.

Tarrow, Sidney G. 2011. *Power in Movement: Social Movements and Contentious Politics*. Cambridge: Cambridge University Press.

Teixeira, Joanison Vicente dos Santos, Jade Silva dos Santos, Deyna Hulda Arêas Guanaes, Wesley Duarte da Rocha, and Alexandre Schiavetti. 2020. "Wild Animals Used as Food Source in the Region of the Serra Do Conduru State Park—PESC, Bahia, Brazil." Preprint. https://doi.org/10.21203/rs.3.rs-88907/v1.

Thayer, Millie. 2010. *Making Transnational Feminism: Rural Women, NGO Activists, and Northern Donors in Brazil*. London and New York: Routledge.

Thomas, William Wayt, Judith Garrison, and Alba L. Arbela. 1998. "Plant Endemism in Two Forests in Southern Bahia, Brazil." *Biodiversity and Conservation* 7 (3): 311–22.

Thompson, James. 2000. "'It Don't Mean a Thing If It Ain't Got That Swing': Some Questions on Participatory Theatre, Evaluation and Impact." *Research in Drama Education: The Journal of Applied Theatre and Performance* 5 (1): 101–4.

Trevisan, Salvador DP. 2019. "A Society-Nature Relationship: The Cocoa Crisis and the Social Movement for Land in Southern Bahia (Brazil) in the 1990s." *Revista de Economia e Sociologia Rural* 36 (3): 193–210.

Trouillot, Michel-Rolph. 1995. *Silencing the Past: Power and the Production of History*. Boston: Beacon Press.

Trouillot, Michel-Rolph. 2001. "The Anthropology of the State in the Age of Globalization: Close Encounters of the Deceptive Kind." *Current Anthropology* 42 (1): 125–38.

Tsing, Anna Lowenhaupt. 1993. *In the Realm of the Diamond Queen: Marginality in an Out-of-the-Way Place*. Princeton, NJ: Princeton University Press.

Tsing, Anna Lowenhaupt. 2011. *Friction: An Ethnography of Global Connection*. Princeton, NJ: Princeton University Press.

Velásquez Runk, Julie. 2017. *Crafting Wounaan Landscapes: Identity, Art, and Environmental Governance in Panama's Darién*. Tucson: University of Arizona Press.

Véran, Jean-François. 2002. "Quilombos and Land Rights in Contemporary Brazil." *Cultural Survival Quarterly* 25 (4): 20–25.

Viatori, Maximilian, and Héctor Andrés Bombiella Medina. 2019. *Coastal Lives: Nature, Capital, and the Struggle for Artisanal Fisheries in Peru*. Tucson: University of Arizona Press.

Vivanco, Luis Antonio. 2006. *Green Encounters: Shaping and Contesting Environmentalism in Rural Costa Rica*. New York: Berghahn Books.

Viveiros de Castro, Eduardo. 2002. "O Nativo Relativo." *Mana* 8 (1): 113–48.

Voeks, Robert A. 1997. *Sacred Leaves of Candomblé: African Magic, Medicine, and Religion in Brazil*. Austin: University of Texas Press.

Walker, Timothy. 2007. "Slave Labor and Chocolate in Brazil: The Culture of Cacao Plantations in Amazonia and Bahia (17th–19th Centuries)." *Food and Foodways* 15 (1–2): 75–106.

Weigand Jr., Ronaldo. 2003. "The Social Context of Participation: Participatory Rural Appraisal (PRA) and the Creation of a Marine Protected Area in Bahia, Brazil." University of Florida. https://ufdc.ufl.edu/UFE0000894/00001.

West, Cornel. 1990. "The New Cultural Politics of Difference." *October* 53: 93–109.

West, Paige. 2006. *Conservation Is Our Government Now: The Politics of Ecology in Papua New Guinea*. Durham, NC: Duke University Press.

West, Paige, and James G. Carrier. 2004. "Ecotourism and Authenticity: Getting Away from It All?" *Current Anthropology* 45 (4): 483–98.

West, Paige, James Igoe, and Dan Brockington. 2006. "Parks and Peoples: The Social Impact of Protected Areas." *Annual Review of Anthropology* 35 (1): 251–77.

Wittman, Hannah. 2009. "Reframing Agrarian Citizenship: Land, Life and Power in Brazil." *Journal of Rural Studies* 25 (1): 120–30.

Wolford, Wendy. 2001. "Case Study: Grassroots-Initiated Land Reform in Brazil: The Rural Landless Workers' Movement." In *Access to Land, Rural Poverty, and Public Action*, edited by Alain de Janvry, Gustavo Gordillo, Elisabeth Sadoulet, and Jean-Philippe Platteau, 304–15. Oxford: Oxford University Press.

Wolford, Wendy. 2004. "Of Land and Labor: Agrarian Reform on the Sugarcane Plantations of Northeast Brazil." *Latin American Perspectives* 31 (2): 147–70.

Wolford, Wendy. 2005. "Agrarian Moral Economies and Neoliberalism in Brazil: Competing Worldviews and the State in the Struggle for Land." *Environment and Planning A* 37 (2): 241–61.

Wolford, Wendy. 2006. "The Difference Ethnography Can Make: Understanding Social Mobilization and Development in the Brazilian Northeast." *Qualitative Sociology* 29 (3): 335–52.

Wolford, Wendy. 2010a. "Participatory Democracy by Default: Land Reform, Social Movements and the State in Brazil." *The Journal of Peasant Studies* 37 (1): 91–109.

Wolford, Wendy. 2010b. *This Land Is Ours Now: Social Mobilization and the Meanings of Land in Brazil*. Durham, NC: Duke University Press.

Woodard, James P. 2005. "Coronelismo in Theory and Practice: Evidence, Analysis, and Argument from São Paulo." *Luso-Brazilian Review* 42 (1): 99–117.

Wright, Angus Lindsay, and Wendy Wolford. 2003. *To Inherit the Earth: The Landless Movement and the Struggle for a New Brazil*. Oakland, CA: Food First Books.

Yeh, Emily. 2013. *Taming Tibet*. Ithaca, NY: Cornell University Press.

Zanotti, Laura, and Natalie Knowles. 2020. "Large Intact Forest Landscapes and Inclusive Conservation: A Political Ecological Perspective." *Journal of Political Ecology* 27 (1): 539–57.

Index

Acre, 228

activism: bottom-up approach to, 244; against coastal development, 184; by GAN'A, 203; growing strength of, 231–32; hunger and, 133–34; land reform, 89–90; networks to organize, 194–97; out in the open, 197; place-based, 194, 203, 205; against Porto Sul, 191–92, 194–97; for quilombos, 117; regional and national, 247; tourism and, 187; by women, 78–79. *See also* social movement actors

Adriana, 132

Adriano, 164

Afro-Brazilian people: as cacao farmers, 42; formal cultural recognition for, 127; inequality in recognition of, 143–44; pride of, 151, 153, 162; on quilombos, 117–18; strategic identities of, 145; ties with quilombos, 115

agency: individual and collective, 98, 113; for quilombolas, 142

Agency for Regional Development, 231

agrarian reform. *See* land reform

agriculture: cabruca, 65–66; cacao, 8, 43–46; colonial, 39–41; diversification, 63, 95; exports from, 188; organic family farms, 62–63; overlaps with conservation, 38. *See also* cacao; family farms/farmers

agroecological movements, 218

agroforestry: cacao production via, 65–66; risk management via, 63; in Southern Bahia, 65

agronomy, 68

Alba, 99, 103, 198, 218, 220, 221

alternative developments, 193

Amado, Jorge, 13, 42, 45, 193

Amazon Forest, 30, 33, 206

analysis: ethnographic, 242–45, 246, 249; new frameworks for, 28; political ecology as context of, 246

Andre, 208

Angélica, 176, 177, 178

Angola, capoeira, 17

animals: Atlantic Forest endemic, 30–32; extinctions of, 212; poaching of, 53, 212; road crossings for, 15

About the Author

Colleen M. Scanlan Lyons is an environmental anthropologist who specializes in sustainable development in tropical forests. She is an associate research professor in the Environmental Studies Program at the University of Colorado Boulder, and director of the Governors' Climate and Forests Task Force. Her work focuses on the Amazon and Atlantic Forests of Brazil, as well as other regions of Latin America, Indonesia, and parts of Africa.